漫论竹炭

Introduction to Bamboo Charcoal

张文标　李文珠　钟金环　编著

Zhang Wenbiao　Li Wenzhu　Zhong Jinhuan

中国林業出版社

图书在版编目（CIP）数据

漫论竹炭 = Introduction to Bamboo Charcoal /张文标，李文珠，钟金环编著. -- 北京：中国林业出版社，2021.10

ISBN 978-7-5219-1200-5

Ⅰ.①漫… Ⅱ.①张… ②李… ③钟… Ⅲ.①竹材—基本知识 Ⅳ.①S781.9

中国版本图书馆CIP数据核字（2021）第186435号

责任编辑： 王思源　马吉萍

电　　话：（010）83143569

出版发行 中国林业出版社 (100009北京市西城区德内大街刘海胡同7号)

http://www.forestry.gov.cn/lycb.html

制　　版 北京五色空间文化传播有限公司

印　　刷 北京中科印刷有限公司

版　　次 2021年10月第1版

印　　次 2021年10月第1次印刷

开　　本 787mm×1092mm　1/16

印　　张 8.25

字　　数 200千字

定　　价 68.00元

《漫论竹炭》编委会

编　　著：张文标　李文珠　钟金环

编委会成员（按笔画顺序排列）：

丁建中　王正郁　王剑勤　毛家女　包立根　朱建华
任敬军　华锡林　刘小康　刘志佳　严王峰　杜松青
吴宗满　吴泉生　汪　芳　张　宏　张　敏　张水祥
张正西　张育民　陈文照　陈冀锐　周家华　郑承烈
涂志龙　程辉武　舒明祥　颜　军　戴美祥

支持单位（按笔画顺序排列）：

上海莱乾新能源有限公司
中国竹产业协会竹炭分会
中国竹炭博物馆
四川惊雷科技股份有限公司
宁波士林工艺品有限公司
竹炭产业国家创新联盟
江山市绿意竹炭有限公司
安吉县华森竹炭制品有限公司
建瓯市恒顺炭业有限公司
浙江农林大学
浙江农林大学生态环境材料研究所
浙江省衢州理工学校
浙江民心生态科技股份有限公司
浙江节节高炭业有限公司
浙江建中竹业科技有限公司
浙江富来森竹炭有限公司
浙江佶竹生物科技有限公司
浙江笙炭控股有限公司
浙江双枪竹木有限公司
遂昌县文照竹炭有限公司
遂昌县神龙谷炭业有限公司
遂昌高净界净化科技有限公司
福建竹家女工贸有限公司
衢州民心炭业有限公司
衢州现代炭业有限公司
衢州净力竹炭科技有限公司
衢州竹韵炭业有限公司

A ROLL OF EDIT COMMITTEE OF
Introduction to Bamboo Charcoal

Authors: Zhang Wenbiao Li Wenzhu Zhong Jinhuan

Author Committee (in alphabetical order):

Bao Ligen Chen Wenzhao Cheng Huiwu Chen Jirui

Dai Meixiang Ding Jianzhong Du Songqing Hua Xilin

Liu Xiaokang Liu Zhijia Mao Jiaonv Ren Jingjun

Shu Mingxiang Tu Zhilong Wang Zhengyu Wu Quansheng

Wang Fang Wang Jianqin Wu Zongman Yan Wangfeng

Yan Jun Zhang Min Zhang Hong Zhang Yumin

Zhang Shuixiang Zhang Zhengxi Zheng Chenglie Zhou Jiahua

Zhu Jianhua

Supporting entities(in alphabetical order):

Anji Huasen Bamboo Charcoal Products Co., Ltd

Bamboo charcoal branch of China Bamboo Industry Associatio

China bamboo charcoal Museum

Fujian Zhujianv Indutry and Trade Co., Ltd

Institute of Ecology Ecomaterial，Zhejiang A&F Univ

Jianou Hengshun Charcoal Industry Co., Ltd

Jiangshan Lvyi Bamboo Charcoal Co., Ltd

National innovation alliance of bamboo charcoal industry

Ningbo Shilin Crafts Co., Ltd.

Quzhou Minxin Charcoal Co., Ltd

Quzhou Xiandai Charcoal Industry Co., Ltd

Quzhou Jinglitan Science and Technology Co., Ltd

Quzhou Zhuyun Charcoal Industry Co., Ltd

Suichang Wenzhao Bamboo Charcoal Co., Ltd

Suichang County Shenlong Valley Charcoal Co. , Ltd.

Shanghai Laiqian New Energy Co., Ltd

Sichuan Jinlei Science and Technology Co., Ltd

Suichang Gaojingjie Purification Technology Co., Ltd.

Zhejiang agriculture and Forestry University

Zhejiang Quzhou Institute of Technology

Zhejiang Minxin Ecological Technology Co., Ltd.

Zhejiang fulaisen bamboo charcoal Co., Ltd

Zhejiang Jianzhong Bamboo Technology Co., Ltd

Zhejiang Jiejiegao Charcoa Co.,Ltd.

Zhejiang Shuangqiang Bamboo Wood Co., Ltd.

Zhejiang Shengtan Holdings Co., Ltd

Zhejiang Jizhu Biotechnology Co., Ltd

序一

竹炭是竹材加工领域利用率最高和附加值最大的科技产品之一。它有较大的比表面积、特殊的孔隙结构，以及优良的吸附能力，可净化室内家具、人造板装修材料、涂料、装饰件等释放的甲醛、苯类、酚类、总挥发性有机物（TVOC）以及空气中的臭气、异味和湿气。同时，竹炭进一步加工后，还具有水质净化、释放红外线、阻隔电磁辐射等诸多功效，在当前大环保与大健康时代，其应用价值越来越突显，如竹炭在气液相环保处理领域作为微生物填料、在土壤修复与改良领域、在农业缓释肥上等均有广泛的应用前景。另外，竹炭可通过深加工成竹质活性炭，其特殊的吸附功能和旺盛的市场需求，更被活性炭界同行所看好，前景十分广阔。

我国竹炭产业在20世纪90年代初期由于门槛较低，主要以作坊式企业为主，存在技术水平、生产和服务能力较弱的问题，产品质量良莠不齐，影响竹炭行业的健康快速发展。经过30多年的技术进步，竹炭生产装备由传统砖土窑型向机械连续化跃升，生产企业由作坊式向规模化转型，产品由单一日用品向功能化深度开发，出现了一批专业化生产竹炭的科技型企业。目前，我国竹炭产量和出口量已跃居世界首位，竹炭加工技术和产品创新能力也处于世界先进水平。其中，固—气联产、电—气和固—气—液联产绿色自动化生产竹炭和竹活性炭技术设备均达到该领域的世界领先水平。

《漫论竹炭》是一本很好的科普读物，有较强的实用性，语言简练，条理清晰，并配以大量的图表和实物照片。全书从竹炭的起源历史开篇，由竹炭的分类、特性和基本知识切入，探索神奇的竹炭世界，领略竹炭的科学、文化、时尚、健康等等。这是作者多年的竹炭教学、科研和产学研结合的杰出成果。本书可为从事竹炭研究、生产和应用者提供专业参考，也可为更广泛的消费者认识竹炭、了解竹炭、使用竹炭提供信息参考。

竹炭产业化实现了竹材增值和循环利用，扩展了竹产业链，完全符合资源消耗减量、循环产出增量的低碳经济发展目标。发展竹炭经济，实践低碳生活，也是应对全球气候变化的有效途径之一。

中国工程院院士

2021年6月

Foreword I

Bamboo charcoal is one of the most value-added products with high material availability in bamboo industry. It presents excellent properties, such as high specific surface area, unique pore structure, and favourable absorption performance, which can be said that bamboo charcoal adsorbs moisture, unpleasant odor and pollutant particles (formaldehyde, VOC, and other toxics) by trapping them while air or water flows through the pores, purifying the surrounding air or water in the process. For that matter, further processing can make bamboo charcoal a perfect candidate material in infrared emission, electromagnetic shielding, and so forth. At a great era of environmental protection and health-care, bamboo charcoal has demonstrated a huge potential in applications in soil amendment and improvement, controlled release fertilizer and microorganism carrier for water management. Meanwhile bamboo charcoal can be activated to prepare bamboo-based activated carbon, a new member to current activated carbon family (coal-/wood-/coconut shell-AC), which is now grabbing serious attention for its powerful performance as well as the renewable raw material supply for the production.

Bamboo charcoal industry in China can be dated back to early 1990s, the production then mainly depended on family artisanal workshop, and production capability, quality level, and technical support are quite low and primitive, which apparently hindered the development of the industry, nevertheless, hundreds of bamboo charcoal(vinegar) products can be available now after a decades of advancement. Traditional brick kilns are replaced by modern automatic mechanical kilns, family workshops have begun to vanish, and large-scale production is becoming a standard practice, functionalized bamboo charcoal products are developed and marketed. The bamboo charcoal industry booms, making China a prosperous bamboo market both in production and trading. The Innovation in bamboo charcoal also pays off, that is, China is at the leading edge of the technology frontier of bamboo charcoal industry, for instance, systemic engineering on co-production of solid(charcoal) and gaseous products, co-production of electricity and gaseous products, co-production of solid, liquid and gaseous products, are industrialized one after another, those works can be versatile to develop the bamboo industry in countries and regions with abundant bamboo resources.

Bamboo charcoal has extend the bamboo industry chain, greatly promoting bamboo recycling, and make value added manufacturing happen, it is in full compliance with the low-carbon development strategy, to put it another way, to develop bamboo charcoal industry is an advantageous approach to reduce costs and increase efficiency for bamboo industry, and it is the very sustainable developmental goals around the world.

Introduction to Bamboo Charcoal, an outstanding work originated from author's decade experiences and expertise in bamboo charcoal research, teaching and industrial practice, introduces informative background, and the production, performance and application concerning bamboo charcoal, which will undoubtedly benefit readers from academy, industry, and customers. Still, this work provides useful explanation and guidance that describes the bamboo charcoal in detail, from which a reader might well be better inspired to handle problems when they arise.

JIANG Jianchun
Institute of Chemical Industry of Forest Products (ICIFP), Chinese Academy of Forestry

序二

竹类植物资源丰富、品种繁多，全球共1642种，面积约5000万hm^2，中国837种，面积641万hm^2。竹与人类文明息息相关，与衣食住行相伴，不仅有良好的涵养水源、生态作用和文化价值，而且有重要的用材价值。21世纪以来，我国竹产业有突飞猛进的发展，竹材产品已经超过万种，在建筑、家居、装饰、交通、包装、造纸、工艺品、碳材料等领域得到广泛应用。

竹炭是竹材高温热解产物，产量高，用途广，已经普遍用于能源、环境治理、化工等领域，为国民经济建设发挥了重要作用。我国竹炭产业自20世纪90年代开始，步入了新的发展阶段，经过30多年的发展，已开发出竹炭、竹醋液300多种产品，2018年我国竹炭产量约为25.52万t，产值约为8.84亿元，但占竹业总产值的份额很小，对竹炭产业来说仍有较大的发展机遇。众所周知，竹炭是利用竹材及其加工剩余物热解得到的固体产物，液体竹醋液、气体竹煤气（可供热、供气和发电），也是竹材加工领域中竹材利用率最高、附加值最大的科技产品之一，在空气净化、水质净化、建筑装修、日化及食品等方面仍然具有重大科技创新空间，逐步引起行业关注。

作者结合多年的竹炭教学、科研和生产实践，总结编写了《漫论竹炭》这本科普读物，从竹炭概述、竹炭原料、竹炭生产、竹炭品质、竹炭应用以及竹醋液生产和应用等方面，深入浅出地展示了竹炭的生产、性能和应用等内容，为从事竹炭领域科技工作者、生产者和销售者提供了专业参考，也为读者更好地认识竹炭、了解竹炭、使用竹炭提供了参考。该书以科普性和实用性为宗旨，语言简练，条理清晰，并配以大量的图表和实物照片，图文并茂，可读性强。

在国家“一带一路”、南南合作等政策指引下，借助国际竹藤组织、国际竹藤标准化技术委员会等平台，推进国际培训和技术推广，使竹炭制造技术和产品不断“走出去”，同时把《漫论竹炭》传播到非洲、东南亚、拉美及加勒比海地区，为世界竹资源高效利用，为改善地球环境、人类健康与产竹区民生经济发展贡献出“中国力量”。

国际竹藤中心研究员 费本华

2021年6月

Foreword II

There are 1,642 documented species of bamboo on the planet with a growing area of 50 m hectares around the world, of the many species of bamboo out there, half of the varieties (837 species) growing in China, with an area of 6.41 m hectares. Bamboo is closely bound up with human civilization, and is quite ubiquitous in all walks of life. It produces great benefits in water conservation, ecological functions, cultural values, and structure materials. The industry in China began to prosper since the 21st century, bamboo related product is everywhere to be seen in construction, home decoration, transportation, packaging, paper-making, crafts, and multiple carbon materials.

Bamboo charcoal is a solid product of bamboo pyrolyzed at a high temperature, which widely applied in energy supply, environmental improvement and chemical engineering, serving the construction of economy in virtue of its excellent properties, which has draw great attention from end users in air and water purification, construction and decoration, and daily chemicals, etc. It is well known that in the process of the carbonization of bamboo materials, gaseous and liquid products can be simultaneously collected and utilized, which suggests the value-adding bamboo charcoal is a most powerful and profitable technological product in bamboo industry. As of 1990s, bamboo charcoal industry entered a new rapid developing era, there are now more than 300 bamboo charcoal(vinegar) products. The bamboo charcoal demand is still thriving and statistics said its yearly production reached about 255 kiloton, generating a 884 million CNY market in 2018.

The authors have a long-term experience in research, teaching, and manufacturing engineering regarding bamboo charcoal, their latest distinguished work, *Introduction to Bamboo Charcoal*, can offer readers a broad industrial developing overview that explained the profound in simple terms concerning raw materials, manufacturing, quality control and application of bamboo charcoal and vinegar. This work also comes complete with selected photos and illustrated instruction sheet. This could readily come in handy for our industrial and professional readers for providing useful guidance in their own work and study.

With the development of the Belt and Road Initiative (BRI), South-South Cooperation(SSC) and the support from International Network for Bamboo and Rattan(INBAR) and Bamboo and rattan Technical Committee of ISO (ISO/TC 296), promotion of bamboo R&D, manufacturing technologies, industrial training and cooperation have been extensively launched worldwide, we hope this brilliant work can make a difference and be beneficial to the rise of world bamboo charcoal industry, which believed to be one of the most effective and efficient route to push forward with global utilization of bamboo resources, especially for these countries and regions with bamboo abundance, improving world environment, and bettering off life for international community of a shared future.

FEI Benhua
International Centre for Bamboo and Rattan(ICBR)

前 言

中国竹资源丰富，竹产业发展迅速，但是产业发展不均衡，竹资源化学利用总体规模小。随着生态文明建设的大力推进和天然林保护工程的深入实施，竹子为改善人类生活、生态环境发挥着积极作用，竹资源利用和产业发展迎来新的发展机遇。

竹子生长快、成材早、产量高、用途广，是一种十分适合加工制造的木质资源。目前竹材的工业化利用主要涵盖了传统竹制品、竹材人造板、竹笋加工品、竹浆造纸、竹炭和竹醋液（含竹叶提取物）等几大类，其中竹炭是竹材及其加工剩余物在缺氧（无氧）条件下的高温热解产物，在森林资源保护、环境、保健以及各种高新技术等领域有着潜在广阔应用前景，对促进社会、经济的可持续发展具有重大意义。

从目前诸多情况来看，竹炭是竹材加工中利用率、附加值较高的利用方式，适宜于各地（国）中小径竹材的开发利用，以及毛竹材制造加工集中区域的废弃物再利用，面对日益激烈的市场竞争，实现竹材的可持续利用发展，亟须重视竹材化学利用，开发竹炭应用技术。

竹炭及其产品经过多年不断发展和深入的开发和应用，已逐渐形成大规模的工业化应用，也进入了人们的日常生活。作者结合多年的教学和科研经验，在竹炭行业发展的基础上，围绕竹炭（含竹醋液等副产物）生产制造、结构性能和应用领域等方面进行了概括与论述，以科普性和实用性为宗旨，语言精练，条理清晰，中英双语并举，以期为国内外从事竹材（化学）利用，尤其是竹炭的研究和生产等相关人员提供专业技术参考，可作为长期在竹炭企业及相关行业从事竹炭材料生产、使用、检测、研究和管理等岗位人员的释疑检索工具、培训教材，也可为对竹炭技术感兴趣的读者提供快速学习帮助。另外，本书还可作为有关院校专业教师和学生的参考资料。

本书由浙江农林大学的张文标、李文珠、钟金环等共同编写、译著，在编写过程中还得到了浙江省重点研发计划项目（2018C02008，2021C03146）、浙江省林业科研成果项目（2017B02）的资助。此外江文正、应伟军、章亮、周海瑛等也参与了本书的部分材料收集与编辑工作。此外，对于书中参阅和引用的相关文献资料，在此向各位研究人员一并表示致敬和谢意。

由于编写人员水平有限，成书仓促，对书中不足之处，敬请读者批评指正。

编著者

2021 年 8 月，于杭州

Preface

China has a long history of bamboo growing, the biomass resource is abundant nationwide, and bamboo industry is booming over past decade, while it develops unevenly, especially the chemical utilization of bamboo materials comes short. With the strategic implementation of ecological civilization and worldwide timber source protection, bamboo industry is of significance in human life better-off and extraordinary opportunity is ready for prosperous development shared by mankind.

Bamboo is a fast-growing *gramineae* plant, featured with high reproduction and wide applications, which covers traditional bamboo works, bamboo-based artificial board, bamboo shoot, pulp and paper-making, and bamboo charcoal (vinegar and other extract inc.), etc, of which bamboo charcoal is a pyrolysis product from bamboo materials in a limited air conditions, it is proved that bamboo charcoal plays an important role in timber source and environment protection, waste management, health care, and other hi-tech fields, promising prospective applications in a sustainable development.

As per current developmental trend, bamboo charcoal is the most value-added, effective and efficient route of utilization of bamboo materials, which in particular is suitable for small diametered bamboo species growing in some region across the world, and bamboo industrial clusters (abundance in processing residue). Extensive attention and innovation shall be ready to meet with fierce competition and challenges from all sides, traditional use of bamboo materials is not enough for the industry.

Herein the authors discussed multiply and ultimately decided to compile an introduction to bamboo charcoal, based on decades of experience in research, teaching and training in bamboo charcoal industry. The authors witnessed the entire developmental history of bamboo charcoal industry from start-up as of 1990s to massive production and application in China today. In this bilingual-published work, it mainly focuses on the development of bamboo charcoal industry, classifications, typical production, fundamental properties and versatile applications of bamboo charcoal and vinegar, with conceptual and conclusive descriptions. It aims to serve people who are now, or will be, involved in R&D, production or marketing in bamboo charcoal industry, used as a helpful reference to solve issues concerning production, analysis, and quality management confronted in actual work and life. The Authors expect individuals and entity from the world participated in bamboo industry can substantially benefit from basic guidance from our humble work.

This work is completed by Zhang Wenbiao, Li Wenzhu and Zhong Jinhuan from ZAFU, Zhong Jinhuan is responsible for the translation, and Jiang Wenzheng, Ying Weijun, Zhang Liang and Zhou Haiying also have contributed to this work. The authors are grateful to the financial support from Key R&D Program (2018C02008, 2021C03146) and *Forestry Achievement Program* (2017B02) of Zhejiang Province. The distinguished works referred in this work, done by our fellow researchers, are also greatly appreciated.

Zhang Wenbiao, Li Wenzhu, Zhong Jinhuan
August, 2021

目 录

Contents

1　竹炭概况

Development of Bamboo Charcoal Industry

1.1 竹炭起源与简史

竹炭，顾名思义，是用竹材烧制而成的炭，比起木炭的使用历史要远远滞后，其生产和应用国家主要有中国、日本、韩国等。

1.1.1 中 国

（1）中国大陆

中国大陆地区的竹炭产业发展大致经历了四个阶段：

第一阶段为萌芽期（1994—2000 年）：

竹炭产业从无到有，生产设备为砖土窑，主要原料为原竹。竹炭生产规模小，产品单一，以出口为主。主要科研单位有浙江林学院（现浙江农林大学）和浙江省林业科学研究院。

1994 年，有炭农开始尝试烧制竹炭。

1995 年，浙江遂昌、衢州、安吉等多地炭农纷纷在传统木炭窑内烧制竹炭。据了解，当时日本市场竹片炭、筒炭销售价格最高时每吨超万元。

1996 年 12 月，成功烧制出第一窑竹炭，掀开了我国竹炭产业新的一页（图 1-1）。

图 1-1 国内早期竹炭试制土窑

Fig 1-1 Brick-soil kiln for early bamboo charcoal production

1.1 Origins and briefs of bamboo charcoal

Bamboo charcoal is literally a charcoal product manufactured from bamboo materials, whose emerging is far behind the history of wood charcoal, and East Asia is the leading region in producing bamboo charcoal (and its products).

1.1.1 Bamboo charcoal in China

(1) Chinese Mainland

There are several stages that can be summarized for China's bamboo charcoal industrial development:

Start-up stage (1994 - 2000):

Bamboo charcoal industry started up from the ground, was produced in brick-soil kilns (Fig 1-1) using bamboo materials on a small scale, the charcoal was then an export-oriented product. Growing together with manufacturers, academic facilities who actively involved in bamboo (charcoal) industrial, like Zhejiang Agriculture and Forestry University (ZAFU), Zhejiang Academy Forestry, and Beijing-based Bamboo Research and Development Center (sponsored by the State Administration of Forestry and Grassland, China), had contributed vastly to the bamboo industry both in China and the rest of the world.

In 1994, the very first plant was set up to produce prototype bamboo charcoal from bamboo materials in a traditional kiln that was formerly used for wood charcoal making in Zhejiang (east China), ever since some informative merchants noted the keen demands in Japan market, the products were exported to

1997年3月，竹炭产品得到日本商户确认，符合出口要求，同年10月开始小批量生产和出口。

1998年，随着国家天然林保护工程重大战略决策的实施，木炭属国家资源类产品，需要消耗大量的天然木材，在这一形势背景下，许多木炭商通过上海华交会和广州的广交会等渠道捕捉到日本市场需要大量竹炭的信息，开始在木炭窑基础上升级改造竹炭窑。同年，浙江省新增竹炭出口产品，产品销往日本、韩国约130t，创汇百万元人民币。

到2000年，浙江省竹炭年出口约1000t。

第二阶段为发展期（2001—2010年）：

竹炭产业逐渐规范，发展迅速，竹炭制备技术进一步提高，产品质量得到明显提升，竹炭产业逐渐辐射至福建、湖南等竹资源丰富产区。竹炭科研单位从几所林业院校发展到浙江大学、清华大学、厦门大学、湖南大学等一批国家重点大学；研究领域也不断扩大，涉及超微竹炭粉加工、竹炭基复合材料、竹炭超级电容器、竹炭食品领域等。

该阶段，共召开了四次竹炭产业发展的国际论坛。2001年10月首届“国际竹炭、竹醋液学术研讨会”在浙江农林大学（原浙江林学院）召开，有来自日本、美国、越南、马来西亚、菲律宾、韩国等国专家学者参加。竹炭知名度逐步提高，掀起了竹炭生产和贸易的热潮，竹炭厂商纷纷开始与科研单位科技合作，促使生产规模进一步扩大，产品加工工艺进一步成熟。2007年6月1—3日，“中国（衢江）竹炭产业发展国际论坛”在浙江衢州举行。

Japan and Korea in 1995, and in 1997 it was certified in Japan. Inevitably many people in the local craft disciplines followed suit in this profitable business, bamboo charcoal industry shaped up naturally. National and provincial forest conservation policy was planned and deployed in the late 1990s, bamboo charcoal as alternative to wood charcoal appeared on the scene.

Growth stage (2001 - 2010):

The bamboo charcoal industry thrived over first decade of 21st century, norms were established, production technologies made progress, quality seemed assured, it charged everyone full-speed ahead, it had spread to neighbouring bamboo regions in China, suck like Fujian and Hunan Province. Thanks to the philosophy of low-carbon green growth, the entities luckily seized the opportunity and restarted afresh. It joined by nation influence universities and institutes, while before which, it was only several regional colleges engaged in the research and development on bamboo (charcoal). Since then, problem oriented research interests arisen from the industry in return became an incentive to its take-off in applications, such as ultrafine powder, novel composite, super capacitor, and colorant in food processing.

During the golden age of bamboo charcoal industry, four (4) grand conferences (see in section 1.2.3) on bamboo charcoal had been hosted which consolidated China's influence in the world as one can see today's giant bamboo charcoal market.

As the financial storm raged in 2008, it had brought extensive challenges about flawed manufacturers. International trade of bamboo charcoal contracted a lot, people reverted to the home market hoping for a rising from the

2009年12月1—2日，“2009中国庆元（国际）竹炭产业发展高峰论坛”在浙江丽水庆元县隆重举行。2010年5月30—31日，“中国竹炭产业基地”建成典礼暨“中国（遂昌）竹炭产业科技创新论坛”在遂昌县隆重举行。

2004年，中华人民共和国商务部、海关总署、国家林业局联合发布公告，对以木材为原料直接烧制的木炭全部禁止出口，竹炭产业政策利好，得到进一步发展。

2008年国际金融危机，竹炭及其制品出口贸易受到严重的冲击，竹炭行业面临新的挑战，竹炭销售逐步转向内销。随着当时低碳环保理念的深入，国内浙江、福建、江西、湖南、四川、广西等地竹炭产业也面临着新的挑战。

第三阶段为转型期（2011—2015年）：

2008年国际金融危机后续影响凸显，加之当时产品主要是日用和初级产品为主，原先不少一哄而上的竹炭企业因资金、管理能力等问题面临洗牌，不少企业因此破产倒闭。此外，竹炭产品过度宣传，功效又难以体验，产业面临困境，迫切需要企业思考如何转型问题。

竹炭产业基地从原来的浙江遂昌、衢州、安吉、庆元和福建的建瓯、永安辐射到浙江宁波、上海、广州和山东等发达地区，竹炭产业起步较早地区呈现出萎缩态势。产品的销售也从原来的团购、连锁专卖等“线下”走向淘宝和“互联网+”的“线上”形态，同时产品结构作了相应调整，从原来简单的竹炭包、竹炭床上用品等扩大到竹炭皮革、竹炭陶瓷砖装饰材料、竹炭基复合材料、竹炭纤维制品及竹炭花

ashes of the exporting dilemma.

Transition stage (2011 - 2015):

Shadowed by global financial tsunami (2008), a plenty of bamboo charcoal manufacturers winded up in bankruptcy, because they overadvertized but disappointedly supplied with primitive, plain bamboo charcoal product to consumers, for that matter, capital deficiency and managemental disorder also accounted for their falldown.

It was urgent for all entities to make a change to move forward. Transition was rather a decision than an option back to that chaotic moment, industrial production center expanded from Zhejiang and Fujian provinces to more locations in China, business model transition happened, familiar group buy and franchise progressively swung to online distribution and marketing in the digital age, and a superb collection of bamboo charcoal products were served: Functional bamboo charcoal leather, decorative panel, fabrics, and other composites. Bamboo charcoal as food colorant is worth to be mentioned from these highlights. Emerging segment markets have grown considerably, say, deodorant (Anji Huasen), air purification (Guangzhou Youke), water purification (Shanghai Yizhu), vehicle seat cushion (Quzhou Xiandai), decorative composite panel (Suichang Biyan), efficient production (Ningbo Xingda), ultrafine powder (Shanghai Hainuo), hookah charcoal (Jiangshan Lvyi), functional fiber (Suichang Mingkang), clothing (Ningbo Guolong), and leather (Suichang Xuanle).

Rising stage (2016 to present):

During this period, substantive transition have been completed, application widened, and production upgraded. Advanced mechanical kilns are employed

生、竹炭饼干、竹炭面条等食用级竹炭产品，不少企业从过去的产品种类繁多、雷同和粗放加工转变为根据企业自身条件开发主导产品，成为更加专业细分市场的生产商，如广州优克室内空气净化用竹炭包，衢州现代炭业汽车专用竹炭垫，宁波国龙功能竹炭布，安吉华森冰箱除味竹炭，遂昌玄乐竹炭皮革，遂昌碧岩竹炭陶瓷砖装饰材料，上海奕竺水质净化用竹片炭，遂昌名康功能型竹炭纤维产品，上海海诺功能性超微竹炭粉、江山绿意水烟炭、浙江旺林食品级竹炭粉、宁波兴达专业生产竹炭原料等等，借此提高了市场竞争力，走出了差异化发展的一步。

第四阶段为升级期（2016 年至今）：

竹炭产业逐渐转型升级，竹炭应用领域不断拓展，产业规模持续扩大，如浙江佶竹、浙江笙炭等公司实现了采用环保装备机械化连续化生产竹炭，年产能可实现上万吨。另外，传统作坊式的砖土窑逐渐被淘汰，开发了一批如竹质活性炭、竹炭装饰板等新产品，应用领域拓展到环境治理等方面；还出现了一些新型颖炭气液联产化自动化制备竹炭的生产企业，一批具有较强引领作用的骨干企业和科技研发中心站在了产业前沿，使产业的整体实力得到了质的提升。

据统计，2018 年我国竹炭产量约为 25.52 万 t，较 2017 年增长 5.67%；产值约为 8.84 亿元，较 2017 年增长 5.87%，辐射带动经济效益 50 多亿元。

（2）中国台湾

陈文祈，被誉为“台湾竹炭发展之父”，中国台湾生态炭产业发展协会秘书

for nonstop manufacturing in companies like *Jizhu* and *Shengtan*, whose yearly throughput could exceed ten thousand (10 000) metric tons, presaging a huge potential in supplanting conventional earth kilns. Eco-friendly application is further expanded, bamboo-based activated carbon, and bamboo charcoal-based composite panels are coming into sight, leading corporations and research facilities at the forefront of the industry always thinks big, even innovative, groundbreaking *charcoal-gas-liquid co-production systems* are already being designed, installed, and operated automatically in those star entities, which paves the way for their solid, lucrative business in forthcoming future.

In 2018, 255200t of bamboo charcoal, with an output value at 884 milion yuan, had been produced, marked 5.87 % rise on a year-on-year basis, a trivial product hereto earned itself a titanic market worth five billion in China Yuan.

(2) Taiwan, China

In 1999, Chen Wenqi, the herald of bamboo charcoal development in Taiwan, return to Taiwan from Japan, where he worked as an apprentice of *Akemi Toba* who respected as Father of Bamboo Charcoal [of Japan], took actions to put what he learned from his mentor into practice, launching the bamboo charcoal business with colleagues from Industrial Technology Research Institute (ITRI).

In 2002, Chen proposed *Transition and Revitalization Programme for Bamboo Industry* to Taiwan's Council of Agriculture, invited Akemi Toba to visit Taiwan and offer institutions to fellow practitioners. Undergrowth was seen then: Production and training bases were initiated by Nantou and

长。1999年，陈文祈任职台湾工业技术研究院（研工院），同年加入日本炭素学会，成为日本“竹炭之父”鸟羽曙的弟子，学习、研究竹炭的烧制和产品开发技术，并将相关竹炭技术和经验带回中国台湾。

2002年，工研院的陈文祈向“农委会”提出“竹产业转型及振兴计划”，并邀请了日本竹炭专家鸟羽曙访问中国台湾并指导产业发展，由此开始形成中国台湾的竹炭产业。随后“农委会”委托中国台湾工业技术研究院研发系列竹炭产品，并由该会林业试验所辅导南投县瑞竹林业合作社及嘉义大埔乡农会，利用台湾优良的孟宗竹、麻竹及桂竹等原料建立了生产教工基地。同年9月在南投县中兴新村举行“921震灾重建三周年”纪念活动，安排了“景观、艺术、文化”的展览，第一批台湾自制竹炭面世。主要产品有竹炭洗发精、沐浴露、洗面奶、美白面膜、竹醋液等5项产品上市；另利用竹炭所生产的高弹性无纺布方面，因质料致密且具弹性、透气轻柔，已完成过滤性精密活性炭纤维原料的开发，应用于工业用粉尘过滤口罩、吸尘器用滤尘袋、空调机用过滤网、滤水器滤心及枕头、椅垫、口罩、眼罩、隔热手套等；还利用竹炭对蔬菜水果所释出的乙烯成分具有良好的吸收能力，研究将活性竹炭涂布或添加于纸浆中，制成竹炭瓦楞纸箱等产品，应用于蔬果包装提高保鲜效果。

2002年，中国台湾工研院计划竹质活性炭方面的研究和应用。

2003年，中国台湾地区建立CAS竹炭产品认证，这是全球第一个竹炭认证标准。符合CAS认证的竹炭产品，其基本条件

Jiayi in central Taiwan. The main bamboo species of the region is: *Phyllostachys edulis* (meso bamboo), *Dendrocalamus latiflorus*, and *phyllostachys bambusoides*. Later in September, 2002, a ceremony on Afterquake Reconstruction marking the 3rd anniversary of *the 921 Earthquake* was hold, first local bamboo charcoal products, containing bamboo charcoal shampoo, body wash, cleanser, beauty mask, and bamboo vinegar, were announced on the *Exhibition of Landscape, Art and Culture*. Bamboo charcoal plants in 2007 increased to 18 as compared to initially 4 factories, but slump is likely to befall in Taiwan's bamboo charcoal industry.

Novel nonwoven fabrics were prepared by incorporating bamboo charcoal in to the polymer, the fabrics is of desired density, flexibility, and breathability, which is a perfect raw material for fabricating dust filter for vacuum cleaner, air conditioner or water strainer, housewares and personal protection equipment.

In 2002, performance and possible use of bamboo-based activated carbon were also investigated by ITRI.

In 2003, Certified agricultural standards (CAS) certification for bamboo charcoal products was advanced, which then was the first accreditation in the world. The suitable raw materials covered meso bamboo grown at least 4 years in pollution-free mountain area in central Taiwan.

At the same year, bamboo charcoal composite fiber was finally developed by Paiho Group, supported by ITRI, in the press the fiber was declared with prospective use in making invisible cloak or garment shunning the detection of night-vision goggles. Paiho also started advertisement campaign for the new product:

是原料必须来自生长在台湾中部海拔无污染山区，且是生长周期四年以上的孟宗竹（毛竹）为主。

2003年，在研工院的技术支持下，台湾百和工业股份有限公司开发竹炭纤维产品，并在工研院的竹炭纤维产品发布会上，发布各类竹炭制服饰和可避开红外线侦测仪的夜间隐形衣，同时成立 LaCoya 品牌。南投鱼池乡的涩水竹炭工作室，突破烧炭技术，用竹子烧出竹炭杯（图 1–2）。

2005年，台湾竹炭纱纺织品上市。同年，陈文祈（图 1–3）在日本爱知博览会"以炭及微生物拯救地球高峰论坛"中以日文发表中国台湾竹炭业经验。

2006年，陈文祈应邀在浙江林学院做《台湾竹炭生产技术现状和发展趋势》的报告（图 1–3）。

2007年，台湾竹炭生产商由最初4家增加至18家。

1.1.2 日 本

1978年就有被称为日本"竹炭之父"鸟羽曙（图 1–4）开始研究竹炭，于1985

图 1–2 竹炭杯
Fig 1-2 Bamboo charcoal mug

LaCoya, a brand means carbon textile in French. At the meantime, bamboo charcoal mug (Fig 1-2) came into being with technical breakthrough in bamboo carbonizing in Nantou's Seshui Bamboo Charcoal Studio.

In 2005, bamboo charcoal fabrics were available in the markets. Chen's (Fig 1-3) report on *Experiencing Bamboo Charcoal in Taiwan*, was addressed on *the Forum on Saving the Planet by Charcoal and Microorganism*, a summit synchronously held and sponsored by *the Expo 2005 Aichi*, Japan. Chen was invited to visit ZAFU and presented a report on development of Taiwan's bambooo charcoal in 2006.

It was said there were 18 bamboo charcoal producers in 2007.

1.1.2 Bamboo charcoal in Japan

As of 1978, Akemi Toba (Fig 1-4), honored as *Father of Bamboo Charcoal* [*of Japan*], begun his scientific exploration on bamboo charcoal, his commercialized bamboo charcoal products was a successful hit on the market in 1985, it shown astonishing performance as compared to binchotan charcoal. The rise of bamboo charcoal quickly earned itself a reputation as *Black Diamond*.

图 1–3 中国台湾竹炭发展奠基人陈文祈报告现场
Fig 1-3 Chen Wenqi, Taiwan's pioneer of bamboo charcoal industry

图 1–4　张文标与鸟羽曙合影（2010）

Fig 1-4　Akemi Toba(L) and Zhang Wenbiao

图 1–5　野村隆哉竹炭技术培训现场

Fig 1-5　Takaya Nomura was giving a technical training on bamboo charcoal, 2004

年成功烧制出竹炭，且将竹炭商业化。由于其性能优越，广受日本大众喜爱，因而又有“黑钻石”之称。日本竹炭和竹醋协会名誉会长、京都大学木质科学研究所野村隆哉在竹炭、竹醋液的生产和利用技术进行了大量的研究工作，并多次访问中国，为中国国内的竹炭、竹醋液发展做出了很大贡献（图 1–5）。

1999 年 5 月，在日本京都大学召开首届世界竹炭、竹醋液学术讨论会，使竹炭产业在日本走向顶峰。近年由于受日本国内经济等因素的影响，竹炭生产和贸易在日本呈现下降趋势。

在 2004 年，日本竹炭的产量为 1566 t，从事竹炭生产人员 886 人，其中爱知县居首，达 117 人。日本和歌山县设有竹炭纪念馆，专门介绍日本竹炭的发展历史、文化和各种应用。

1.1.3　韩　国

韩国有着悠久使用木炭的历史，但韩国竹炭的发展较迟，1997 年才开始生产竹

Takaya Nomura, former chairmen of Japanese Bamboo Charcoal and Bamboo Vinegar Association, had explored a lot in production and application of bamboo charcoal and vinegar, made multiple trips to China and gave lectures, which helped boost China's bamboo industrial growth (Fig 1-5).

In May, 1999, the first International Symposium on Bamboo Charcoal and Bamboo Vinegar was hold in Kyoto University, Japan. Bamboo charcoal flourished, at present, both the production and trade of bamboo charcoal are going through a rough patch because of the economic downturn over past decades.

In 2004, there was a yield of 1 566 tons of bamboo charcoal in Japan with 886 individuals involved in the industrial, of which there were 117 in Aichi-ken of central Japan. There is a Bamboo Charcoal Monument in Wakayama of south-central Japan, focusing on development, culture, and applications of bamboo charcoal.

1.1.3　Bamboo charcoal in Korea

There is a long history of using charcoal in Korea, for instance, *Dae Jang Geum,* royal

炭，以机械制炭方式为主，由韩国森林科学院提供技术支持。

20世纪90年代末，韩国木炭研究所姜在允所长对木炭（包括竹炭）有广泛研究（图1-6）。2003、2005年分别出版了《木炭拯救性命——徐徐揭开的秘密》韩文、中文版。

据木炭研究所相关人员介绍，在2005年，韩国开始从中国大量进口竹炭。2008年有规模的生产商2家，年总产量约为170t。竹炭主要用于烧酒生产（韩国大众白酒著名品牌“真露”）、烟草过滤嘴生产和水净水、空气净化、织布染色、土壤改良以及作为各种燃料等。

韩国林产物生产统计并无竹炭一项，目前木炭占据主导地位。竹炭未能产业化，且作为竹炭主原料的竹林未能得到及时培

图1-6 张文标（左）与姜在允（中）合影（2019）
Fig 1-6 Kang Jae-yoon (M) and Zhang Wenbaio(L)

chef of *Jewel in the Palace*, a hot Korean TV series with story background in 1480s of the Korea Dynasty, once used a charcoal as deodorant in a soy sauce vat for a purpose of stink removal, yet the modern development of bamboo charcoal lagged behind the follow countries and regions until a mechanical kiln was used to prepare bamboo charcoal in 1997, which was technically supported by Korea Forestry Promotion Institute.

Kang Jae-yoon (Fig 1-6), the direct of Korea Charcoal Research Institute, started the research on (bamboo) charcoal in the late 1990s, and Wood Charcoal saving life, -- an Unveiled Secret was also published in Korean and Chinese languages in 2003 and 2005, respectively.

According to the Institute's briefing, Bamboo charcoal, mostly imported from China since 2005, were used in alcohol distillation of brewing, tobacco filters, water and air purification, cloth dyeing, soil amendment, and fuel applications. In 2008, there were two factories in scale production with a total yield of 170 metric tons annually.

Industrialization of bamboo charcoal production is under development with difficulty in replacing wood charcoal, which then was not included in Korea's forest products statistics, either.

Bamboo forest was not fully cultivated in time, and low cost of bamboo products from other east and southeast Asia made local bamboo material economically unprofitable and uncompetitive, in fact, the bamboo forest area in Korea is decreasing year by year, even unfortunately, the Authorities did not concern and no industrial policy agenda encouraged. However, some individuals and entities were aware of bamboo forest can be served as

育。由于成本低廉的竹产品从中国及东南亚的大量进口，培育竹林并无经济效益和市场竞争力，所以韩国国内的竹林处于逐年减少的境况。而且，目前韩国政府也未采取特别的倾斜和培育政策。但是，有部分地区和企业将竹林视为观光休养地，并以此为目的培育竹林。譬如：竹乡（全罗南岛潭阳郡，系历史悠久的竹乡，类似中国的安吉）为了营造更多的观光休养竹林，正在投放人力和物力大力扩大竹林面积。

1.2 竹炭组织与交流

1.2.1 协 会

（1）浙江省竹产业协会竹炭分会

①协会成立：浙江省竹产业协会竹炭分会，2001 年 9 月在浙江林学院召开，共 40 余家从事竹炭生产、销售、科研和教育等企事业单位和个人会员单位参加成立大会。日本竹炭协会名誉会长、日本京都大学木质材料研究所野村隆哉教授，日本京都大学木质材料研究所张敏博士应邀参加会议。大会选举产生了分会会长、副会长和理事单位，通过了分会章程。

②机构设置：秘书处设在浙江林学院。第一届浙江省竹产业协会竹炭分会会长：张齐生院士；秘书长：叶良明；副秘书长：张文标。

2007 年 10 月—2017 年 9 月，浙江省竹产业协会竹炭分会会长：张齐生；秘书长：张文标；副秘书长：张宏。

2017 年 9 月—2019 年 5 月，浙江省竹产业协会竹炭分会秘书长：张文标；副秘书长：张宏。

fashional relaxation and leisure destination, and bamboo was grown. In Damyang-gun County of Jeolla Island, a reputable bamboo town, great efforts have been conducted to bamboo cultivation for building a tourist destination.

1.2 Organizations and exchanges

1.2.1 Associations in China

(1) Bamboo Charcoal Branch of Zhejiang Bamboo Industry Association

The Branch was founded in Sep, 2001 with the Secretariat in Zhejiang Agriculture and Forestry University, it has more than 40 founding members covering production and marketing entities, education and research institutes, individual members are also welcomed. The aim of the Branch is to served local enterprises in Zhejiang Province, where there is most developed bamboo industrial in China (Fig 1-7).

(2) Bamboo Charcoal Branch of China Bamboo Industry Association

The Branch was lately founded in May, 2019, in Zhejiang Agriculture and Forestry University. On the founding day, a series of International Standards Organization (ISO) standard drafts concerning bamboo charcoal had been fully discussed word by word and an agreed draft successfully coordinated by over 100 experts of different countries, including representatives from academia and industrial community. The NPO develops its membership from academia and industrial whom involves in bamboo charcoal (and vinegar) products manufacturing, research, teaching, and consulting across the country.

（2）中国竹产业协会竹炭分会

①协会成立：中国竹产业协会竹炭分会于 2019 年 5 月 17 日成立，大会在浙江农林大学举行。中国竹产业协会常务理事长刘志佳，以及来自全国的竹炭企业代表、相关行业专家和参加 ISO/TC296 WG3《竹炭》国际标准制定研讨会的国内外专家共 100 余人出席此次成立大会。大会根据要求选举产生了分会理事长、副理事长和常务理事单位，通过了分会管理办法（图 1–7）。

②机构设置：秘书处设在浙江农林大学。第一届中国竹产业协会竹炭分会理事长张文标；秘书长：张宏；副秘书长：马中青，马建锋。

图 1–7　中国竹产业协会竹炭分会成立大会（2019）

Fig 1-7　The foundation of the Bamboo Charcoal Branch of China Bamboo Industry Association, 2019

1.2.2　联　盟

2019 年成立了林业和草原国家创新联盟——竹炭产业国家创新联盟，由国际竹藤中心牵头，浙江农林大学、浙江大学、福建农林大学、中国林业科学研究院林产化学工业研究所、浙江省林业科学研究院及若干竹炭规上企业组成的联盟，其目的是充分发挥联盟在协同创新、服务林业现代化建设中的作用，为推动竹炭产业高质量发展做出更大贡献。

1.2.2　Industrial alliances in China

National Innovation Alliance of Bamboo Charcoal Industry was co-founded in 2019, by universities, institutes, and enterprises. The main parties included: the International Center of for Bamboo and Rattan (ICBR, China), Zhejiang Agriculture and Forestry University, Zhejiang University, Fujian Agriculture and Forestry University, Institute of Chemical Industry of Forestry Products (ICIFP, China Foresty Academy), Zhejiang Foresty Academy and multiple bamboo charcoal enterprises, aiming to collaborative innovation, serving the forestry modernization and facilitating the quality development of bamboo charcoal industry.

1.2.3　会　议

（1）（中国）国际竹炭、竹醋液学术研讨会（2001）

2001 年 8 月，首届“2001（中国）国际竹炭、竹醋液学术研讨会”在浙江林学院举办（图 1–8），应邀参加会议的国内外专家和企业家包括日本竹炭竹醋液协会会长野村隆哉，日本九州竹炭竹醋液协会会

1.2.3　Conferences in China

(1) (China) International Symposium on Bamboo Charcoal and Bamboo Vinegar, 2001

Held in Hangzhou (capital city of Zhejiang, a province with absolutely largest number of bamboo charcoal producer, and developed industry in China), on Aug, 2001 (Fig 1-8).

The symposium on bamboo charcoal that

图 1–8 （中国）国际竹炭、竹醋液学术研讨会（2001）

Fig 1-8 International Symposium on Bamboo Charcoal and Bamboo Vinegar (2001)

长末广胜也，日本京都大学木质材料研究所张敏博士，以及来自美国、越南、马来西亚、菲律宾、韩国的多名专家学者。

会议期间，国内外专家在会上发表演讲并指出：

①在完善竹炭生产工艺方面，要建立竹材含水率控制和窑体温度自控系统；分析不同培育方式和不同竹龄的竹材对热解工艺与产品性能的影响，研制新型竹炭窑；开发纯氧制炭技术，探讨增加热解副产品回收率的新途径；

②在产品研究和开发方面，要进一步扩大竹炭和竹醋液的应用领域；研制开发竹炭复合板和炭纸等功能型材料；利用竹炭的电导率特性，开发新型半导体材料，应用于电子和航空技术领域；开展竹醋液在医药领域的应用研究，使竹醋液能应用于治疗糖尿病和湿疹等疾病。

会议出版了会议论文集《竹炭竹醋液机能科学》，参会代表考察了浙江省遂昌县、衢江区两地的竹炭企业。此次会议标志着中国竹炭企业真正意义上的崛起。

（2）中国（衢江）竹炭产业发展国际论坛，2007

held since the loosen industrial was readily to march forward, was a milestone standing for the rapid rise of the bamboo charcoal cause in China.

Experts and professional producers of bamboo charcoal who were in meeting came from Japan, Korea, Malysia, the philppine, the United States, and Viet Nam. Science and technical researches were reported, which conerned the most common and key difficulties and breakthrough that confronted in production and application, e.g. moisture content monitering, in-kiln temperature control, the effect of bamboo (speices, growth age, etc) and pyrolyisi process on charcoal quality and performance, equipments and apparatus for production and quality control, by-product recycling, high value applications in pharmaceutical, semiconductor, composite and electronics. To be brief, since then the industry began to soar.

A Proceedings was published too. Local bamboo charcoal plants were also visited by attendees.

(2) Forum on China's Bamboo Charcoal Industrial Development (Qujiang), 2007

Held in Quzhou (southwest Zhejiang), on June 1-3, 2007 (Fig 1-9).

2007年6月1—3日，“2007中国（衢江）竹炭产业发展国际论坛”在浙江省衢州市举行（图1–9）。论坛由浙江省林业厅和衢州市衢江区人民政府主办，浙江省竹产业协会、衢州市衢江区林业局承办，浙江林学院等单位协办。论坛的宗旨是进一步增进国际间竹炭产业的交流与合作，解决竹炭产业发展中遇到的难题，完善竹炭产业发展规划，明确今后发展方向，探讨竹炭产品开发思路，开拓国内外市场，提升中国竹炭知名度。

参加本次论坛的有中国工程院院士、浙江林学院院长张齐生，纳米材料研究专家、中国科学院院士都有为，竹炭吸附、净化功能研究专家、中国林科院林产化工研究所研究员邓先伦，浙江林业厅副厅长邢最荣，浙江林学院副院长、教授方伟，日本竹炭协会副会长阿部良幸，日本京都大学生态环境材料研究所张敏教授，韩国树林产业社长姜德中，衢州市委常委、副市长雷长林，衢州市林业局局长徐土土，中共衢州市衢江区委书记毛建民，浙江省竹产业协会竹炭分会秘书长、浙江林学院工程学院叶良明教授、浙江省竹产业协会竹炭分会副秘书长张文标博士，此外还有韩国堡力士公司、日本通商株式会社等代表出席了会议。

论坛上张齐生院士作《竹炭的神奇功能》、都有为院士作《竹炭纳米材料开发与应用》、阿部良幸会长作《日本竹炭竹醋液行业现状及其标准》、姜德中社长作《韩国竹炭竹醋液产品现状及前景》、邓先伦研究员作《竹质颗粒活性炭在公共场所的净化功能和节能》等报告。

图1–9 中国（衢江）竹炭产业发展国际论坛，2007
Fig 1-9 Forum on China's Bamboo Charcoal Industrial Development , 2007

The forum was cohosted by Zhejiang Forestry Adminstration and local Qujiang governmet, it aimed to seek mutual exchange, to tackle challenges in an innovative and collaborative way, to establish credibility and market exposure for bamboo charcoal products in everyone's interest.

Zhang Qisheng (Chinese Academy of Engineering), Du Youwei (Chinese Academy of Science), Deng Xianlun (Chinese Academy of Forestry), Yoshiyuki Abe (Bamboo Charcoal Association of Japan), and Kang De Jung (Korea Society of Forestry Industry) lectured on development and prospects of bamboo charcoal (vinegar) in the Forum.

Other experts and delegates were from Zhejiang Agriculture and Forestry University, Bamboo Charcoal Branch of Zhejiang Bamboo Industy Association, Kyoto University (Japan), and provincial and local civil servants of Zhejiang, and quite a number of industrial representatives had attended the forum. The forum was exposured with considerable coverage in the national and local media.

(3) Summit of Bamboo Charcoal Industrial Development (Qingyuan), 2009

《CCTV-7 中国农民致富报道组》《浙江日报》《科技信息报》《浙江科技报》《衢州日报》衢江区新闻部等新闻媒体也高度关注此次盛会，并派记者参加了此次盛会。

（3）中国庆元（国际）竹炭产业发展高峰论坛，2009

2009 年 12 月 1—2 日，“2009 中国庆元（国际）竹炭产业发展高峰论坛”在浙江省丽水市庆元县隆重举行（图 1-10）。本次高峰论坛由省林业厅、浙江林学院、南京林业大学、丽水市人民政府联合主办，浙江省竹产业协会竹炭分会和庆元县人民政府共同承办。活动以“加强技术合作交流、促进竹炭产业发展”为主题，旨在通过国际学术研究活动，加强竹炭产业的合作与交流，拓宽竹炭产业市场，促进竹产业整体发展进程，扩大庆元竹产业的影响力。

中国工程院院士、浙江林学院名誉校长张齐生教授，国际竹藤组织高级项目官员傅金和出席论坛。浙江省林业厅副厅长刑最荣、浙江林学院副校长鲍滨福、中国竹产业协会秘书长程美瑾分别在开幕式上致辞。

国内外知名专家、竹业主产区政府领导及竹制品企业负责人等共计 180 余人参加了高峰论坛。国外专家有日本生物质炭素材料研究会顾问，京都大学生存圈研究所张敏教授，日本竹炭、竹醋液生产者协议会顾问，名古屋大学木村志郎教授，日本竹炭、竹醋液生产者协议会会长鸟羽曙，加拿大阿尔伯特研究院生物质炭研究所汪孙国研究员，加拿大阿尔伯塔省农业和农村发展农业研究分区尼古拉·萨维多夫博

Held in Qingyuan (a county of Lishui, southwest Zhejiang), on Dec.1-2, 2009 (Fig 1-10).

The summit was coorganzied by Zhejiang Forestry Adminstration, Zhejiang Agriculture and Forestry University, Nanjing Forestry University, and Lishui Municipal government, it aimed to enhance cooperation and exchange for a better development of bamboo charcoal industry, it also helped the Quzhou with grown influence in the industry.

Zhang Qisheng (Chinese Academy of Engineering), Fu Jinhe(INBAR) and Xing Zuirong (Deputy Director-General Zhejiang Forestry Adminstration), Bao Bingfu (vice-president of Zhejiang Agriculture and Forestry University), and Chen Meijin (secretary general of China Bamboo Industry Association) delivered speches on their congratulations.

More than 180 industrial experts, business representative government employees attended the summit. Foreign invited guests from University of Alberta (Canada), Universität Freiburg (Germany), University of Limerick (Ireland), Bamboo Charcoal and Bamboo

图 1-10　中国庆元（国际）竹炭产业发展高峰论坛，2009

Fig 1-10　Summit of Bamboo Charcoal Industrial Development, 2009

士，爱尔兰利默里克大学罗伯特·弗拉纳根博士，韩国木炭研究所姜在允所长，德国弗莱堡大学西格玛·舍恩赫尔博士等。

论坛期间，中国工程院院士张齐生教授作了《生物质材料的新利用技术与途径》的主题报告。国际竹藤组织高级项目官员傅金和作《国际竹炭发展现状及技术需求》，全国竹藤标委会秘书长张禹作《竹炭标准化建设及标准申报程序》，日本京都大学生存圈研究所研究员张敏、日本竹炭竹醋液生产者协议会顾问作《日本的竹炭和竹醋液产业的现状》，韩国木炭研究所姜在允作《韩国竹炭竹醋液产业现状和发展前景》，名古屋大学名誉教授木村志郎作《竹材和竹炭的高效切削加工》，加拿大尼古拉·萨维多夫博士作《生物质炭——未来生长的媒介》，罗伯特·弗拉纳根博士作《竹炭生物质炭在土壤中应用研究》，浙江省竹产业协会竹炭分会秘书长张文标博士作《中国竹炭产业现状和发展趋势分析》等报告，各专家学者和企业界代表进行了广泛交流。

（4）中国竹炭产业科技创新论坛，2010

“中国竹炭产业基地”建成典礼暨“中国竹炭产业科技创新论坛”于2010年5月30—31日在遂昌县隆重举行。本次会议由中国竹产业协会和浙江省林业厅联合主办，浙江省竹产业协会和浙江省遂昌县人民政府共同承办，该论坛以“发展竹炭经济，引导低碳生活”为主题，旨在转变经济发展方式，倡导低碳生活，“以竹代木”生态效益巨大，发展前景广阔。国家林业局领导和浙江省林业领域的代表共100余人参加了会议。全国政协人口资源环境委员会副主任、国际竹藤组织董事会联合主席、

Vinegar Producer Association (Japan), Kyoto University (Japan), Nagoya University (Japan), and Korea Wood Charcoal Institute (Korea) also had their work addressed in the Summit.

During the forum, Professor Zhang Qisheng, academician of the Chinese Academy of Engineering, as well as domestic and foreign experts and scholars made a report on the production and application of bamboo charcoal and bamboo vinegar, while representatives of the enterprises and experts conducted extensive exchanges and discussions.

(4) Forum on Innovation of China’s Bamboo Charcoal Industrial, 2010

Held in Suichang (a county in Lishui, Zhejiang), on May 30-31, 2010.

The Forum was co-hosted by Zhejiang Forestry Administration, Zhejiang Bamboo Industry Association and Suichang County government, themed with Developing bamboo charcoal economy, leading a low carbon life. Leaders of Population, Resources and Environment Committee of the CPPCC, ICBR, CBIA attended the meeting and delivered speeches (Fig 1-11).

Presentations on bamboo charcoal industrial development were reported by experts from INBAR, Japan, Korea, and China. The topic contained R&D and applications of both bamboo charcoal and vinegar around the world, and after the Forum, the ground breaking ceremony for Bamboo Charcoal Industrial Base (Suichang country was accredited earlier) was witnessed by attendees, later experts and representatives visited China Bamboo Charcoal Museum which was highly complimented, and hoped the developing of the industrial can lead a sustainable growth for entire green economy.

中国竹产业协会会长江泽慧教授出席会议并致辞（图 1–11）。

会议举行了中国竹炭产业基地建成典礼仪式，江泽慧会长代表中国竹产业协会授予浙江省遂昌县和福建省建瓯市“中国竹炭产业基地”的称号。会议举办了“中国竹炭产业科技创新专家报告会”，国际竹藤网络中心常务副主任岳永德教授主持报告会，日本株式会社野村研究所所长野村隆哉，韩国木炭研究所所长姜在允，国际竹藤组织项目官员傅金和等来自国内外的 7 名专家学者和 2 家企业代表作了有关竹炭和竹醋液的科技开发、功能评价和产业发展等方面的专题报告。会议期间，与会代表还参观了世界第一个竹炭博物馆，对我国竹炭产业取得的成就给予高度评价，并相信中国竹炭产业今后会取得更大的发展，为低碳经济做出贡献。

（5）竹炭产业高峰论坛，2017

2017 年 6 月 5—6 日，竹炭产业高峰论坛暨浙江省竹产业协会竹炭分会年会在浙江省丽水市召开（图 1–12）。会议由浙江省林业厅、丽水市人民政府主办，浙江省竹产业协会、浙江省竹产业协会竹炭分会、丽水市人力资源和社会保障局、丽水市林业局承办，会议还邀请了国内外专家、企业家等 80 多人，其中有日本九州竹炭竹醋液协会会长末广胜也、日本雄本县木竹炭振兴会五通洋二部长，日本京都大学生存圈研究所张敏教授等。

会议期间，邀请末广胜也会长作报告《日本竹产业现状和发展趋势》，浙江富来森中竹科技有限公司陈再华技术总监作报告《竹炭竹醋液产业转型升级》，纳爱斯

图 1–11　江泽慧在中国竹炭产业基地建成典礼上发言，2010

Fig 1-11　Jiang Zehui, Chairperson of CBIA, gave a speech on commissioning ceremony of China Bamboo Charcoal Industrial Center, 2010

图 1–12　竹炭产业高峰论坛会场，2017

Fig 1-12　Bamboo Charcoal Industrial Summit, 2017

(5) Bamboo Charcoal Industrial Summit, 2017

Held in Lishui, on June 5-6, 2017 (Fig 1-12).

Annual Conference of Bamboo Charcoal Branch (of Zhejiang Bamboo Industry Association) was held at the same time in Lishui, hosted by Zhejiang Forestry Administration and Lishui Municipal government. Academic and industrial experts, from abroad and home, were invited.

Bamboo charcoal and its development in Japan was introduced, and the fellows, bamboo charcoal (vinegar) producer and user, promoted their product, which drew

集团有限公司研发中心段建军主任作报告《竹炭竹醋液的日化产品应用》，浙江农林大学、浙江省竹产业协会竹炭分会张文标博士、教授分别作《竹炭》国际标准项目（ISO21626）及《专用竹片炭》林业行业标准制订情况、浙江省竹产业协会竹炭分会年度工作汇报，与会代表对竹炭产业发展和标准的制定进行热烈的讨论提出了许多建设性意见。

（6）竹炭竹醋液生产与应用研讨会，2017

2017 年 12 月 12—14 日在浙江省湖州市安吉召开竹炭竹醋液生产与应用研讨会，浙江省竹产业协会竹炭分会，浙江农林大学生态环境材料研究所主办，浙江佶竹生物科技有限公司承办，邀请日本和国内竹炭专家、学者、企业、省林业厅、安吉县林业局相关领导共 52 人参加（图 1–13）。

论坛期间邀请日本奈良炭化工业株式会社社长玉川甲泰作《竹木醋液和竹木炭在日本的生产和利用最新状况》，浙江大学生物系统工程与食品科学学院教授盛奎川作《竹材生物质材料的热解工艺及产品性能》，国际竹藤中心研究员汤锋作《竹醋液对农药的增效作用及复配制剂研制》，广州薇美姿实业有限公司技术总监陈敏珊作《竹炭竹醋液在洗涤护肤品上应用》，纳爱斯集团公司有限公司研发中心主任段建军作《竹炭竹醋液洗涤产品研发研究报告》，日本京都大学生存圈研究所渡边所长助手李瑞波博士作《竹醋液对口蹄疫的预防效果研究报告》，南京农业大学教授李春梅作《竹醋液在畜牧养殖上的应用研究》，江西师范大学产业设计研究所教授江新喜作《食品级竹醋开发应用研究》等报告。

extensive attention. Prof. Zhang Wenbiao (the author) also submitted the annual bulletin to the Branch, also announed the ISO standard proposal on bamboo charcoal to the audience on the Summit, suggestions and opinions on the draft were well received.

(6) Symposium on Production and Application of Bamboo Charcoal and Bamboo Vinegar, 2017

Held in Anji county (south Zhejiang), on Dec 12-14, 2017 (Fig 1-13).

The symposium cohosted by Bamboo Charcoal Branch of Zhejiang Bamboo Industry Association, Zhejiang Agriculture & Forestry University, and Zhejiang Jizhu Biotechnology Co,Ltd. Experts from Nara Carbonization Industry (Japan), Zhejiang University (China), ICBR, Kyoto Univeristy(Japan), Nanjing Agricultural University (China), Jiangxi Normal University (China) and renowned entities in bamboo charcoal and vinegar applications, gave technical detailed reports on their recent advances in cosmetics, feed additive, food processing.

(7) Bamboo charcoal producers, Huasen, and Jizhu were visted, whose kiln was brick-earth and mechanical style, respectively

Global Bamboo and Rattan Congress, BARC 2018

图 1–13　竹炭竹醋液生产与应用研讨会，2017

Fig 1-13　Symposium on Production and Application of Bamboo Charcoal and Bamboo Vinegar, 2017

会议期间，与会代表还参观了安吉县华森竹炭制品有限公司砖土窑竹炭生产基地，并在浙江信竹生物科技有限公司并进行了座谈。

（7）世界竹藤大会竹炭产业发展和竹炭标准研讨会，2018

2018 年 6 月 25—27 日，首届世界竹藤大会在中国北京召开，来自世界各国竹子、竹炭方面专家应邀参加会议（图 1–14），会议另设立一个分会场，主题是：竹炭产业发展和竹炭标准研讨，由国际竹藤组织，ISO/TC 296 竹炭专家委员会主办，国际竹藤中心，浙江省竹产业协会竹炭分会承办，张文标教授主持，来自菲律宾、加纳、乌干达、美国、哥伦比亚、印度尼西亚、韩国、日本、老挝、越南等各国专家学者对《竹炭》ISO 国际标准的内容进行广泛的讨论。

（8）ISO/TC 296《竹炭》WG3 工作组国际标准制定研讨会，2019

2019 年 5 月 15—16 日在浙江农林大学召开《ISO/TC 296〈竹炭〉WG3 工作组国际标准制定研讨会》（图 1–15）。会议由竹藤技术委员会（ISO/TC 296）、国际竹藤

Held in Beijing, in June, 2018 (Fig 1-14).

A subconference on Bamboo Charcoal Standard and Industrial Development was arranged by INBAR, cohosted by ISO/TC 296 Bamboo and Rattan Committee, ICBR, and Bamboo Charcoal Branch of Zhejiang Bamboo Industry Association, and Prof. Zhang Wenbiao (the author) chaired the subconference. Experts from Colombia, Ghana, Indonesia, Japan, Laos, the Philippines, South Korea, Uganda, the United States, and Viet Nam, attended and discussed the working draft on bamboo charcoal (ISO/WD 21626).

(8) Symposium on Bamboo Charcoal Standard Preparation, 2019

Held in Hangzhou, on May 15-16, 2019 (Fig 1-15).

The symposium was co-organized by ISO/TC 296, Zhejiang Agriculture & Forestry University, and the International Center of for Bamboo and Rattan (ICBR, China). The working group delegates from Colombia, Indonesia, Kenya, Nepal, Nigeria, the Philippines, and Uganda, and guest experts from Kyoto University (Japan), Zhejiang University (China),

图 1–14 世界竹藤大会竹炭产业发展和竹炭标准研讨会，2018

Fig 1-14 Global Bamboo and Rattan Congress, BARC 2018

图 1–15 ISO/TC296《竹炭》WG3 工作组国际标准制定研讨会，2019

Fig 1-15 Symposium on Bamboo Charcoal ISO Standard Preparation, 2019

中心主办，浙江农林大学、浙江省竹产业协会竹炭分会承办。来自乌干达戴维·凯胡尔、印度尼西亚古斯坦·帕里、菲律宾杜尔塞布兰卡·蓬萨兰和瑞秋·珍妮·皮迪克、尼日利亚约瑟夫·乌巴贾、哥伦比亚胡利安·乌兰塔多、肯尼亚保罗·穆尼、尼泊尔苏希尔·库马尔·沙阿等国外专家、ISO/TC 296秘书处、全国竹藤技术标委会刘贤淼秘书长、国际竹藤中心陈绪和研究员、日本京都大学张敏教授、浙江大学盛奎川教授、南京林业大学周建斌教授、西南林业大学王慷林教授、竹炭企业代表等参会，会议对标准进行细致认真研讨，形成ISO《竹炭》国际标准草案。会后，还到安吉县参观浙江佶竹生物科技有限公司，了解机械化竹炭和竹活性炭生产情况，并对竹炭产业方面相关问题进行广泛交流。

1.3 竹炭之乡与基地

1.3.1 竹炭之乡

2001年浙江省衢江区被国家林业局命名为全国首个“中国竹炭之乡”。

2008年遂昌县被国家林业局经济林协会授予“中国竹炭之乡”称号（图1–16）。

2012年衢州被国家林业局经济林协会授予“中国竹炭之乡”称号。

1.3.2 产业基地

浙江省遂昌县、浙江省衢州市、福建省建瓯市先后被授予“中国竹炭产业基地”称号。

2010年5月30日中国竹产业协会授予浙江省遂昌县和福建省建瓯市“中国竹炭

Nanjing Forestry University (China), Southwest Forestry University (China), and various enterprises, attended the meeting, discussed and closed with fully inspected draft, which later was submitted in plenary meeting in Septerber in Manila, Philippine in September, 2019. Delegates and experts visited and exchanged the novel briquette as well as automatic production in Zhejiang Jizhu Biotechnology Co. Ltd, the top supplier of bamboo-based activated carbon in China.

1.3 Industrial clusters and geographical indications

1.3.1 Town of Bamboo charcoal

In 2001, Qu-jiang District, of Qu-zhou City (in southwest Zhejiang, East China) was authenticated as first Town of Bamboo Charcoal by then State Forestry Bureau (now the State Administration of Forestry and Grassland, China).

Sui-chang County (neighbouring to Qu-zhou) and Qu-zhou City were named as the Town of Bamboo Charcoal in 2008 and 2012, respectively (Fig 1-16).

1.3.2 Industrial production centers

There are several Bamboo charcoal industrial centers in China: Suichang County, Quzhou City, and Jianou City (in Fujian province neighbouring to southwest Zhejiang).

In May, 2010, Suichang County and Jianou City were announced at the same time as *Bamboo Charcoal Industrial Center* in China at the opening ceremony of Bamboo Charcoal Museum (Fig 1-17). And

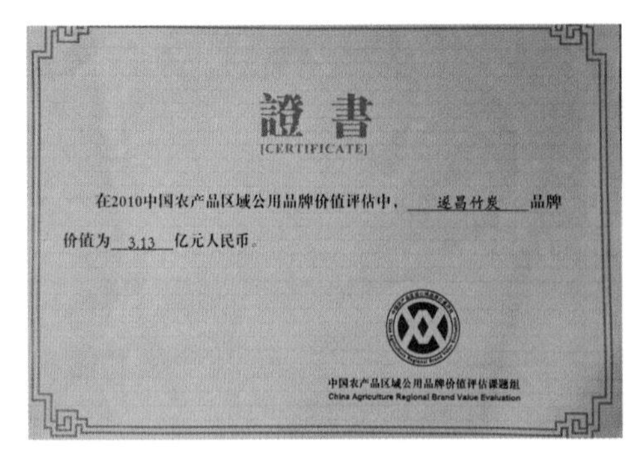

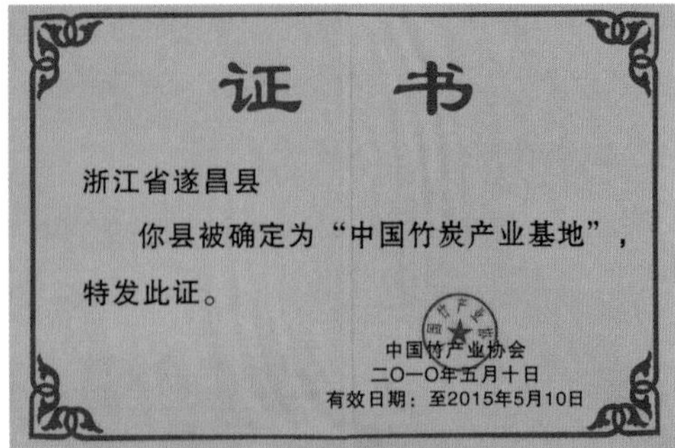

图 1–16 竹炭之乡

Fig 1-16 Town of Bamboo charcoal

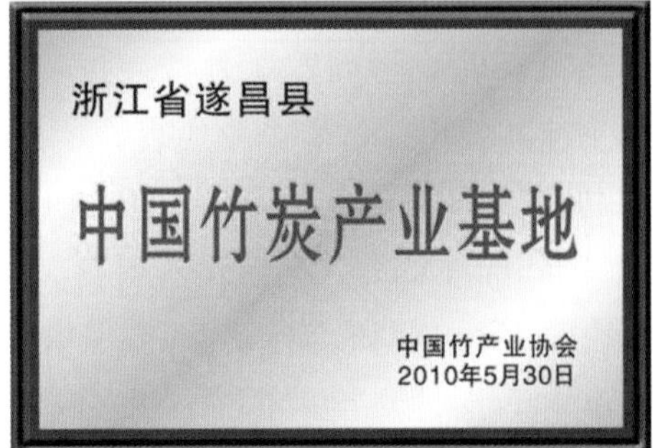

图 1–17 中国竹炭产业基地

Fig 1-17 Bamboo charcoal industrial center

产业基地”的称号（图 1–17）。全国政协人口资源环境委员会副主任、国际竹藤组织董事会联合主席、中国竹产业协会会长江泽慧，省政协副主席冯明光，浙江省林业厅厅长楼国华等出席了在遂昌县中国竹炭博物馆举行的“中国竹炭产业基地”建成典礼。来自国家林业局，中国竹产业协会，浙江省、市、县的相关负责人，国内外的专家学者、生产企业、销售客商及新闻媒体共 200 余人参加了此次活动。

2012 年衢州被授予“中国竹炭产业基地”的称号。

1.3.3 地理标志

在 2006 年，“遂昌竹炭”成为中国地

again in 2012, Quzhou City was named as *Bamboo Charcoal Industrial Center.*

1.3.3 Geographical indications

In 2006, established and renowned Suichang bamboo charcoal was officially registered as Product of Geographical Indication of China, soon afterwards China's national standard, GB/T 21819—2008 Product of geographical indication - Sui-chang bamboo charcoal, was issued and implemented by the State Administration for Market Regulation (SAMR), which have contributed substantially to the fast development of bamboo charcoal products in China.

理标志产品，随后GB/T 21819—2008《地理标志产品 遂昌竹炭》正式发布并执行，在提升竹炭特色产品质量、促进区域经济发展和对外贸易等方面将发挥着越来越大的作用。

1.4 竹炭标准与贸易

1.4.1 标 准

（1）中国

①国家标准

GB/T 21819—2008《地理标志产品 遂昌竹炭》

GB/T 26913—2011《竹炭》

GB/T 26900—2011《空气净化用竹炭》

GB/T 28669—2012《燃料用竹炭》

GB/T 30365—2013《寝具竹炭》

GB/T 30124—2013《竹炭涤纶低弹丝》

GB/T 30125—2013《竹炭涤纶短纤维》

GB/T 31744—2015《水质净化用竹炭基本性能试验方法》

②行业标准

LY/T 1929—2010《竹炭基本物理化学性能试验方法》

LY/T 2144—2013《空气净化用竹炭包》

LY/T 2221—2013《竹炭生产技术规程》

LY/T 2483—2015《竹炭产品术语》

LY/T 3203—2020《竹炭远红外发射率测定方法》

LY/T 3205—2020《专用竹片炭》

FZ/T 52014—2011《竹炭粘胶短纤维》

FZ/T 64079—2020 面膜用竹炭粘胶纤维非织造布

1.4 Industrial standards and trading

1.4.1 Industrial standards

Bamboo charcoal industrial is most developed since 1990s in China, there are a number of standards have been issued, and more is being drafted.

China's national standards (GB), forestry standards (LY), and textile standards (FZ):

GB/T 21819—2008 *Product of geographical indication - Sui-chang bamboo charcoal*;

GB/T 26900—2011 *Bamboo charcoal for air-purification;*

GB/T 26913—2011 *Bamboo charcoal;*

GB/T 28669—2012 *Bamboo charcoal for fuel;*

GB/T 30124—2013 *Bamboo-charcoal polyester drawn textured yarn;*

GB/T 30125—2013 *Bamboo charcoal polyester staple fiber;*

GB/T 30365—2013 *Bamboo charcoal for home furnishings;*

GB/T 31734—2015 *Bamboo Vinegar*

GB/T 31744—2015 *Test on the elementary properties of bamboo charcoal for water-purification*;

LY/T 1929—2010 *Test on the elementary physical and chemical properties of bamboo charcoal;*

LY/T 2144—2013 *Bamboo charcoal prepackage for air-purification ;*

LY/T 2221—2013 *The producing technical regulation of bamboo charcoal;*

LY/T 2483—2015 *Terminology of bamboo charcoal products;*

LY/T 3203—2020 *Measurement method for far infrared emissivity of bamboo charcoal;*

（2）国外

日本竹炭竹醋液生产者协议会（平成17年3月，2005年）《日本竹炭分类标准》。

牙买加 DJS 333—2014《空气净化用竹炭》。

国际标准化组织的竹炭相关 ISO 系列标准由浙江农林大学张文标教授（项目领导）主导制订：

ISO 21626—1：2020 竹炭第1部分：《竹炭的通用要求》；

ISO 21626—2：2020 竹炭第2部分：《燃料用竹炭》；

ISO 21626—3：2020 竹炭第3部分：《净化用竹炭》；

牙买加：DJS 333—2014 空气净化用竹炭技术条件；

日本竹炭和竹醋液生产商协会：木醋液和竹醋液标准。

1.4.2 世界竹炭产业分布

通过国际竹藤组织、国际竹藤中心和国家林业和草原局竹子研究和开发中心等部门举行的各种竹子培训班，以及中国实施的“一带一路”倡议，至今竹炭产业已经扩展到亚洲、非洲、南美洲等国家和地区，具体如下。

亚洲：日本、中国、韩国、菲律宾、印尼、马来西亚、印度、越南、孟加拉国、缅甸、泰国等；

非洲：埃塞俄比亚、加纳、肯尼亚、莫桑比克、坦桑尼亚、马达加斯加等；

南美洲：牙买加、哥伦比亚、厄瓜多尔、智利等；

此外，在法国、俄罗斯、加拿大、新

LY/T 3205—2020 *Special bamboo charcoal flake;*

FZ/T 52014—2011 *Bamboo charcoal vscose staple fiber;*

Other standards:

ISO 21626—1:2020 *Bamboo charcoal - Part 1: Generality;*

ISO 21626—2:2020 *Bamboo charcoal - Part 2: Fuel applications;*

ISO 21626—3:2020 *Bamboo charcoal - Part 3: Purificiation applications;*

Jamaica*: DJS 333-2014 Specification for Bamboo Charcoal for Air-purification*

Japan*: Producer Association of Japanese Bamboo Charcoal and Bamboo Vinegar* (Heisei March, 2005);

1.4.2 Global industrial and trading

A wide variety of training course on bamboo and bamboo products have been co-hosted by *the International Network for Bamboo and Rattan* (INBAR, an NPO founded in 1997), *the International Center of for Bamboo and Rattan* (ICBR, China), and *Bamboo Research and Development Center* (the State Administration of Forestry and Grassland, China). Since the launch of the Belt and Road Initiative in Sep. 2013, bamboo charcoal industrial have been moving into the fast lane of development over continents with abundant bamboo resources:

Asia: China, Japan, South Korea, the Philippines, Indonesia, Malaysia, India, Vietnam, Bangladesh, Myanmar, Thailand,etc.;

Africa : Ethiopia, Ghana, Kenya, Mozambique, Tanzania, Madagascar, etc.; and,

South America : Jamaica, Colombia, Ecuador, Chile, etc.

Still, bamboo charcoal trade is happening in Europe and America.

加坡等国也有竹炭贸易。

1.5 文化传承与发展

1.5.1 竹炭博物馆

2008年，遂昌县文照竹炭有限公司投资2000万元，建立了总用地面积16600平方米，总建筑面积6万多平方米的中国首个竹炭博物馆。

中国竹炭博物馆是国内首家以炭历史文化及国内外炭产品展示为主题的博物馆，博物馆有炭祖大殿、炭文化历史展、炭综合应用展馆、炭科学原理体验馆（青少年科普中心）、炭缘客栈、竹炭美食、炭窑酒吧、炭养生馆、炭旅游休闲购物一条街。

竹炭博物馆并不是传统概念里的博物馆，而是以“观、吃、住、娱、购”为一体的旅游景点，实现企业由二产的竹炭加工向三产服务业的转型升级。

1.5.2 非物质文化遗产

遂昌“竹炭烧制技艺”于2009年被列入第三批浙江省非物质文化遗产名录（图1-18）。烧木炭是我国的千年传统技艺，而烧竹炭是对这项工艺的坚守与创新，通过非物质文化遗产的薪火相传，竹炭产业必将更加蓬勃兴旺。

1.5.3 竹炭专业职业教育

2009年，衢江区职业中专（现浙江省衢州理工学校）与浙江民心炭业股份有限公司等企业合作开办了竹炭工艺专业。为了提高教学质量，2013年衢江区职业中专成立了由

1.5 Bamboo charcoal cultural development

1.5.1 Bamboo Charcoal Museum

China Bamboo Charcoal Museum, with an area of 16 600 m^2 (total floor area: 60 000 m^2), had been built with a budget of 20 million China Yuan (2.88 million USD) in 2008 by *Sui-chang Wen-zhao Bamboo Charcoal Co., Ltd.* It focuses on charcoal culture exhibition and global charcoal products in the Museum. There are several showrooms and sections in the Museum: Hall of the Forefather of Charcoal, Hall of Charcoal History and Culture, Hall of Charcoal Applications, Visitor Center (Experience and Science Education Center), Taiyuan Inn, Charcoal Food Plaza, Kiln Bar, and Charcoal Health-care Hub, and leisure shopping area. The Museum is now a popular local tourist attraction, serving people with diversified (bamboo) charcoal culture, featured products, and relaxation and from stressed work.

1.5.2 Intangible Cultural Heritage

In 2009, *the Maneuver of Bamboo Charcoal Making* in Sui-chang (southwest Zhejiang, of East China) was registered in *the Intangible Cultural Heritage List of Zhejiang (third editon)* (Fig 1-18).

1.5.3 Vocational education

Qu-jiang Vocational School, located in Qu-zhou (southwest Zhejiang), set up a new major: Bamboo Charcoal, with assistance from Zhejiang Agriculature and Forestry University (ZAFU) and *Zhejiang Min-xin Charcoal Co, Ltd.* The vocational education project offers various training course for

图 1–18　中国竹炭博物馆（遂昌）

Fig 1-18　China Bamboo charcoal Museum

图 1–19　竹炭“产学研”研讨及竹炭教学

Fig 1-19　Authors' involvement in vocational education on bamboo charcoal

高校、企业、中职学校组成的竹炭专业“产学研”联合体，同年该联合体被评定为市级“产学研”联合体，2013 年 8 月该专业被评为浙江省特色专业（图 1–19）。

students who will be a future participant in bamboo charcoal industrial, the collaborative achievement was proudly appraised as one of *Zhejiang Featured Education Project* in 2013 for which serves the industrial, promoting the steady development of local bamboo charcoal industrial (Fig 1-19).

2 竹炭类别

Classification of Bamboo Charcoal

2.1 炭与碳

2.1.1 炭的来历

“炭”字古已有之。东汉许慎的《说文解字》(大徐本)中就说:“炭,烧木余也。”《说文》的另一版本(小徐本)则称:“炭,烧木未灰也。”描述得更准确,更科学。《周礼·月令》上也说,“草木黄落,乃伐薪为炭”。唐代诗人白居易的名篇《卖炭翁》中有“卖炭翁,伐薪烧炭南山中”,更加清晰地指出炭就是由薪材烧制而成的。在我国古代,炭就是木炭,这毫无疑问。“炭”字也衍生出很多转义,如“炭,火也。”(《玉篇》);“炭,墨也。”(《孟子·公孙丑上》);汉代还有以炭为姓的人(《西京杂记》),最重要的一个转义是,炭也指煤炭,如《正字通》:“炭,石炭,今西北所烧之煤。”日本仍有把煤称作石炭的,发音也和汉语基本相同。自古以来,炭便是一个常用字,不管是天然的炭,还是人造的炭,在我国一直是用不带“石”字旁的炭来命名的,这种用法至少已有两千多年的历史。

当今的炭是指一种既传统又现代的材料,其作为燃料已被人们普遍认知。炭还是一种多孔性材料,具有很强的吸附、调湿、净化、除臭等功效。目前常见的有煤炭、炭黑、焦炭、木炭、活性炭、成型炭(机制炭、烧烤炭)、竹炭及生物质炭等。

2.1.2 碳

“碳”字,《说文》上没有,号称收字

2.1 Charcoal and carbon

2.1.1 Historical background of charcoal

Charcoal (Chinese character: 炭 , tàn) was recorded in an ancient Chinese dictionary: *Word and Expression* (*Shuo Wen,* by *Xu Shen*, 58 - 147 AD, of the Han Dynasty), according to explanation edition by *Xu Xuan* (916-991 AD, of the Song Dynasty), “*charcoal is the residue of burned wood”*, but according to the edition by *Xu Kai* (920 - 974 AD), “*charcoal is the residue of burned wood, yet not ashed”* (Note: Xu Xuan is elder brother of Xu Kai).

In *Rites of Zhou* (*Anonymous*, published in the Han dynasty: 202 BC - 220 AD. The Zhou Dynasty: 1046 BC - 256 BC), *“when grasses and leaves turn yellow, trees can be cut down for making charcoals.*”

Great poet *Bai Ju-yi* (772 - 846 AD, of the Tang Dynasty) wrote in his masterwork, *The Old Charcoal Seller*, “*The old charcoal seller is cutting firewood to make charcoal in Nanshan mountain...*”. All those literature works are suggesting what material can be used to prepare (wood) charcoal.

A lot of meanings were deprived from charcoal, for instance, “charcoal is flame” (*Jade Dictionary,* or *Yupian*, by *Gu Ye-wang*, 519 - 581 AD), “charcoal is ink” (*The Works of Mencius* by *Mencius*, 372 BC - 289 BC, of the Warring State period), and so forth. In the Han Dynasty, some practitioners who prided in the product and proudly using *Charcoal* as surname (from *Miscellany of the Western Capital*, by *Liu Xian,* 50 BC - 23 AD, of the Han Dynasty). In an ancient dictionary, *Knowledge of Characters (Zheng Zi Tong,*

最全的《康熙字典》上也没有，近代编辑的《中华大字典》上没有，1983年出版的《辞源》增补本上仍然没有。20世纪20年代末、30年代初出版的词书上，开始出现带“石”字旁的“碳”字。如1930年出版的《王云五大字典》就收录了这个字，指明其唯一意义是符号为C的元素，元素周期表中第6号化学元素。“碳”字的诞生是1932年11月26日。

2.1.3 炭与碳

在很多文献和场合，带“石”字旁的“碳”与不带“石”字旁的“炭”常被混用，可谓“炭碳不分”。炭 = 碳（无定形碳或石墨）+ 有机物 + 无机物 + 水分。碳是一种化学元素，有多种同素异形体，如无定形碳、石墨（六方晶系）及金刚石（立方晶系），凡涉及化学元素C的名词均用“碳”，如碳键、碳链、碳环、碳化钙、固定碳等。炭是以碳为主并包含多杂质的混合物，如煤炭、焦炭、炭黑、活性炭、木炭、竹炭等。工业上含碳的制品应称炭制品，如炭砖、炭电极、炭块、炭刷、炭纤维、炭布、炭毯等。

2.2 炭分类

2.2.1 煤　炭

煤炭是古代植物埋藏在地下经历了复杂的生物化学和物理化学变化逐渐形成的固体可燃性矿物。煤炭主要含有碳（C）、氢（H）、氧（O）、氮（N）和硫（S）等元素，还有极少量的磷（P）、氟（F）、

by Zhang Zi-lie, 1597 - 1673 AD, of the late Ming Dynasty), it recorded: *"Stone charcoal is coal wildly used in burning stove in northwest China".* Still today somewhere in Japan, coal is called stone charcoal, whose pronunciation is almost the same with that in China.

From ancient times to the present, natural or man-made charcoal that commonly used in daily life, is a traditional and modern carbonaceous material well known as fuel, moreover, the porous structured charcoal shows great potential in absorption performance in current industrial and home applications.

2.1.2 Carbon

Chinese character (碳 , tàn), equivalent to *Carbon,* was not existed in the dictionary: *Word and Expression (Shuo Wen)*, even cannot be found in *the Kangxi Dictionary*, a dictionary compiled (in 1710 - 1716 AD) during the reign of Kang Xi in the Qing Dynasty with most Chinese characters recorded by far. It was also not expected in *Grand dictionary of Chinese characters* (published in 1915), and *Chinese Etymology Dictionary* (in 1983). But it firstly appeared in *Grand Dictionary of Wang Yun-wu* (published in 1930s), indicating that its meaning is the chemical element with the symbol C and atomic number 6. In conclusion, Chinese character (碳 , tàn) did not come into existence until Nov., 1932.

Note: Chinese character (炭 , tàn) is equivalent to charcoal, and character (碳 , tàn) is equivalent to Carbon, they have identical pronunciation, but judging from the characters' configuration, the discrepancy between them is only one stone (石 , shí).

2.1.3 Charcoal and carbon

Charcoal, technically is a complex

氯（Cl）和砷（As）等元素。碳（C）、氢（H）、氧（O）是煤炭有机质的主体，占95%以上；煤化程度越深，碳的含量越高，氢和氧的含量越低。煤炭根据用途可分为动力煤和炼焦煤；按煤炭的挥发分含量多少，将煤分为褐煤、烟煤和无烟煤。

2.2.2 焦　炭

烟煤在隔绝空气的条件下，加热到950~1050℃，经过干燥、热解、熔融、黏结、固化、收缩等阶段最终制成焦炭。焦炭主要用于高炉炼铁和铜、铅、锌、钛、锑等有色金属的鼓风炉冶炼，起还原剂、发热剂和料柱骨架的作用。

2.2.3 炭　黑

炭黑是由气态或液态的碳氢化合物（主要为石油）在空气不足的条件下进行不完全燃烧或热裂分解所生成的微细黑色粉末。按用途分为橡胶用炭黑和色素用炭黑。炭黑主要组成物是碳元素，还含有少量的氢、氧、硫、灰分、焦油和水分，是工业中不可缺少的化工原料，是仅次于钛白粉的重要颜料，具有染色与补强的特性，是多数塑料、橡胶制品的改质添加剂。

2.2.4 木　炭

木材经过炭化后，剩余的固体物质即为木炭；燃烧时会不时发出噼啪声或产生火花，由于燃烧过程未经人工化学处理，因此其主要成分为碳及少量灰分。国家标准（GB/T 17664—1999）中定义木炭有硬阔叶木炭、阔叶木炭和松木炭。其中硬阔叶木炭是由硬阔叶材及桦木材的混合材烧

mixture of carbon (amorphous carbon), organic compounds, inorganic compounds, and water (as its moisture content). Coal, one of the fossil fuels, contains not only amorphous carbon, but also a variety of chemicals (or so-called impurity).

Carbon is a nonmetallic chemical element with atomic number 6 that readily forms compounds with many other elements forming an enormous number of important molecules, many of which are essential for life. There are several allotropes of carbon, e.g. amorphous carbon, graphite, diamond, fullerene (also known as footballene), carbon nanotube, graphene, and new carbon phases may as well be about to unveil.

2.2 Classification of charcoals

2.2.1 Coal

Coal is a combustible black or brownish-black sedimentary rock (stone) with a high amount of carbon and hydrocarbons. Coal is classified as a nonrenewable energy source because it takes millions of years to form. Coal contains the energy stored by plants that lived hundreds of millions of years ago in swampy forests. It mainly consists of carbon (C), hydrogen (H), oxygen (O), nitrogen (N) and sulfur (S), (approx. 95% by weight in total), and phosphorus (P), fluorine (F) and chlorine (Cl) are also found in trace concentrations. It can be classified as thermal and coking coal according to applications, and Lignite, bituminous and anthracite coal according to volatile matter.

2.2.2 Coke

Coke is a fuel with few impurities and a

制的炭。阔叶木炭是由硬、软阔叶的混合材烧制的炭。松木炭由松木或针叶材烧制的炭。

木炭出口贸易在我国已被禁止。贸易上木炭又分为白炭和黑炭，以白炭贸易为主。两种木炭主要在原材料、炭化工艺、产品理化性能及用途上有差异。

白炭是指木材（主要是乌岗木类）在窑内炭化完毕后趁热取出，盖上湿沙土，使木炭在熄灭过程中与空气接触而进行煅烧，表面被氧化成白色。白炭质地坚硬，比重大，断面有金属光泽，敲击会发出清脆的钢音，燃烧时不发烟，发热量高，燃烧持久，无黑色粉末散落，适用于水质净化。黑炭是木材（软阔叶材类）在窑内炭化完毕后即闷窑熄火得到。其较松软，比重小，易着火，燃烧时常冒烟并伴有爆裂声，发热量低，不耐烧，易散落黑色粉末，适用于吸湿和除异味（图 2–1）。

2.2.5 备长炭

备长炭是日本的一种木炭，其外观上有一层泛银白的灰色，炭化温度在 1000℃以上，燃烧时火力稳定而持久，在燃料、饮食、生活日用方面有着广泛应用。关于备长炭由来有以下几种说法。

一种说法是，备长炭（乌冈炭），据记载为中国（皇帝炭）传到朝鲜后由朝鲜又

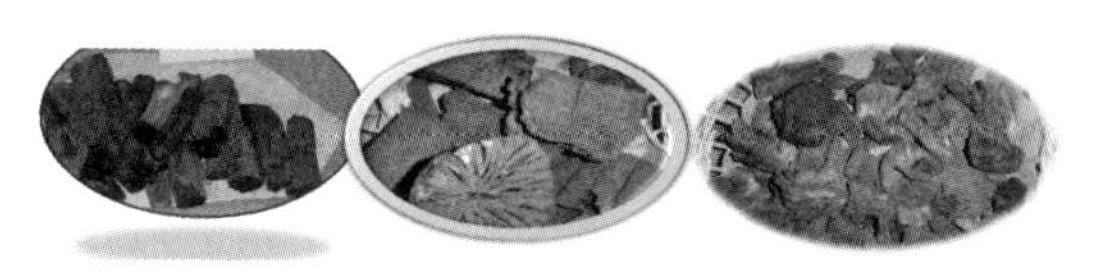

图 2–1 白炭（左）、黑炭（中）和松木炭（右）
Fig 2-1 White charcoal(L), black charcoal(M) and pine charcoal (R)

high carbon content, usually made from coal. It is the solid carbonaceous material derived from destructive distillation in the absence of air, from low-ash and low-sulphur bituminous coal, at temperatures usually around 950-1 050℃ . The process vaporizes or decomposes organic compounds in the coal, driving away volatile matters, including water, in the form of coal gas and coal tar. The non-volatile residue of the decomposition is mainly carbon, in the form of a hard, slight glass-like solid that cements together the original coal particles and minerals.

It is an important industrial raw material, used in iron ore smelting, the raw iron is made by reducing (removing the oxygen from) iron ore (iron oxide) by reacting it at high temperature with coke in a blast furnace.

2.2.3 Carbon black

Carbon black is produced with the thermal decomposition or the partial combustion using hydrocarbons such as oil or natural gas as raw material in little oxygen condition or no air available.

The characteristics of carbon black vary depending on manufacturing process, and therefore carbon black is classified by manufacturing process. Carbon black produced with the furnace process, which is the most commonly used method now, is called "furnace black". Carbon black is one most important and applied chemicals next to titanium dioxide in pigment market, wildly used in rubbers and plastics industrial as colorant and enhancement materials.

2.2.4 Wood charcoal

Wood charcoal is a porous, highly carbonaceous product formed during the heating of wood without access (or with

传到日本（备长炭）。

另一种说法是，大约在 18 世纪初，一位名叫备中屋长左卫门的日本商人，为标榜自己制作的木炭有特色，取自己名字中的“备”和“长”，合成商店名，开了一家名为“备长”的木炭店，于是，他经手的木炭就称为“备长炭”。从此，这个“牌子”流行起来，这就是备长炭的起源。

再有种说法，在日本江户时代的元禄年间，位于现今的和歌山县田边市的木炭批发商备中屋长左卫门从 1854—1973 年历经 120 年全力普及纪州白炭，后世人就采用其名字的略称，把纪州白炭称作备长炭。

备长炭原料的主要树种为:（日本）姥芽栎 / 马木坚木 / 姥目坚 / 橡木（*Quercus phillyraeoides*），（中国）乌擦棒 / 壳斗科栎属坚木 / 乌岗栎 / 乌刚栎。

备长炭近年成为白炭的代名词，亦有在日本国外如中国等地生产。市场上出现备长炭，其实应该写作“备长炭”，“備”和“备”是同音字。

2.2.6 活性炭

活性炭是指生物质材料经过活化具有活性的物质（图 2–2）。它是一种用途很广

图 2–2 竹质活性炭

Fig 2-2 Bamboo-based activated carbon

limited access) to air in furnaces and retorts (sometimes in campfires as well).

White charcoal, a hardwood charcoal, is the main product in trade, and black charcoal is ordinary charcoal having quite a big share in the market, they differ in raw material, carbonization process, product properties and end applications. The quality of the charcoal differs mostly depending on how the fire is extinguished.

White charcoal is made by carbonizing the wood at a moderately low temperature, then, near the end of the process, the kiln temperature is raised to approximately 1000℃ to make the wood red hot. When making white charcoal, it needs to be quite skilled in removing the charcoals, which have turned deep red, from the kiln and quickly smother it with a covering of powder (say, sandy soil or ash) to cool it. This will then give a whitish color to the surface of the charcoal, from which the name "white charcoal" was derived (Fig 2-1).

According to the manufacturing, the quick rise in temperature followed by quick cooling, burns up the outer layer of the wood leaving a smooth hardened surface, as hard as cast iron. That is also the reason a white charcoal called a “hard charcoal”. When knocked, metallic clanking sounds will be heard from the white charcoal. As a premium product, white charcoal may take some more time to ignite, but its thermal conductivity is way better than ordinary black charcoal. The smoke-free flame produced evenly by natural white charcoal provides more energy, lasts long enough to be used as a fuel of choice to cooking, due to its quick heating with no explosive cracking during combustion. In addition, white charcoal are regularly applied

的吸附剂和催化剂，是由一些含碳的原料（如木材、竹子、果壳、秸秆、煤、沥青、焦炭等）经过炭化和活化等加工过程生产出来的。根据外观形状可分为粉末状、颗粒状及柱状等；根据使用原料可分为植物类（木屑、木炭等）、矿物类（煤、石油和焦炭等）及其他（纸浆、水解木素、废塑料等）；按制造方法可分为气体活化法活性炭、化学药品活化法活性炭及混合活化法活性炭；按用途可分为气相吸附用（如溶剂回收活性炭、脱硫炭等）、液相吸附用（如糖用炭、味精炭和水处理炭等）及催化剂或催化剂载体用活性炭等。

2.2.7 竹　炭

竹炭是竹材在高温缺氧或限制性通氧气的条件下经炭化得到的黑色固态产品（图 2-3）。竹炭通常选用 4 年生（含）以上的竹材经过干燥、预炭化、炭化和煅烧四个阶段制成的。竹材炭化过程实质上是竹材细胞壁化学成分纤维素、半纤维素和木质素等热分解反应的总和。

2.2.8 成型竹炭

成型竹炭通常又称机制竹炭、烧烤竹

图 2-3　竹炭
Fig 2-3　Bamboo charcoal

in water purification.

Generally, to make black charcoal the wood is carbonized (partial burning with little air) at temperatures ranged from 400℃ to 700℃ , then the kiln is sealed until the burning stops and the heat slowly extinguishes. The surface of the charcoal is black, soft and retains the outer layer of the wood. It is also easy to ignite and burns hot enough (sometimes giving explosive cracking noise) that it was used as fuel for ordinary daily food cooking including industrial use during the former times, instead, it can be used for deodorizing and humidity conditioning, which occasionally called “domestic charcoal”.

The wood materials for carbonization to produce charcoal is now confronting with rigid challenges from timber resource protection, in many countries and regions, wood charcoal trading is strictly regulated, even forbidden, which makes eco-friendly bamboo charcoal an excellent alternative to the charcoal trading.

2.2.5 Binchotan charcoal

Binchotan charcoal, also known as white charcoal, is a very pure high carbon charcoal made from oak (*Quercus phillyraeoides*).

According to historic record, binchotan charcoal, originated from China as (emperor charcoal), firstly begun to circulate in Korean, and later spread to Japan and got its current name.

Some people also say, there was a famous charcoal artisan, Binchoya Chozaemon, over 300 years ago in the Kishu province of Wakayama in Japan, who thought highly of his exclusive charcoal products and proudly named it as binchotan charcoal. His secret binchotan process eventually became known in other parts of Japan, but Kishu Binchotan

炭。通常有 A、B 两种类型，A 型是竹粉经干燥，挤压成型经炭化而制得的固体炭（图 2-4）。该类产品主要有六角形中心有孔、四方形中心有孔和多边形中心有孔等多种类型。整个生产过程中未添加任何胶黏剂，它是传统木炭的替代产品，具有燃烧时间长、热值高、不冒烟、不发爆、环保等多种优点，常用于工业领域。

B 型是以竹炭粉为主要原料，添加胶黏剂进行混合，经成型干燥等工序而制得的固体。常用的胶黏剂有淀粉胶、羟甲基纤维素（CMC）等。同 A 型竹炭相比，存在燃烧时间短、热值低、会冒烟等缺点，常用于家庭日用领域。

图 2-4　成型竹炭（上 A 型，下 B 型）

Fig 2-4　Bamboo charcoal briquettes, BCB-a (top) and BCB-b(bottom)

remains the finest, densest and purest form comparing to competitors, until now it is still handmade through a unique process passed down through generations. Another say is that, Binchoya Chozaemon, a wholesaler and distributor, worked all out to promote his white charcoal, which afterwards was called Binchotan charcoal in memory of him.

Binchotan charcoal outperforms regular charcoal products in hardness, density and fixed carbon content, which is considered as high quality charcoal in the market.

2.2.6　Activated carbon

Activated carbon (also called activated charcoal, or active carbon) is a carbonaceous (Fig 2-2), highly porous adsorptive medium that has a complex structure composed primarily of carbon atoms, featuring high surface area, pore structure (micro-, meso- and macro-), and high degree of surface reactivity, activated carbon is a widely applied adsorbent and catalyst carrier, to purify, dechlorinate, deodorize and decolorize both liquid and vapor applications, or to be modified(loaded) with catalyst to enhance its performance in use.

There are basically two methods for preparing activated carbon: physical and chemical activation. Physical activation consists of two steps: the carbonization of the starting (raw) material, and the activation of the char by using carbon dioxide or steam. In chemical activation both the carbonization and the activation step proceed simultaneously. The raw materials for making activated carbon can be biomass (wood, coconut, bamboo, crop straws, etc), and fuel-based products (coal, tar, coke, etc.)

2.2.9 生物质炭

国际生物质炭协会将生物质炭（biochar）定义为生物质材料在限制性通入氧气的环境条件下通过热化学转化而制得的固体材料。另外一种定义指生物质高温裂解的固体产物，一般为具有高度芳香化、富含碳素的多孔固体颗粒物质，其含有大量的碳和植物营养物质、具有丰富的孔隙结构、较大的比表面积且表面含有较多的含氧活性基团，是一种多功能材料。它不仅可以改良土壤、增加肥力，吸附土壤或污水中的重金属及有机污染物，而且对碳氮具有较好的固定作用，施加于土壤中，可以减少二氧化碳（CO_2）、氮氧化合物（NO_x）、甲烷（CH_4）等温室气体的排放，减缓全球变暖。

2.3 竹炭分类

2.3.1 形状分类法（图 2–5）

筒炭：筒状的竹炭，用外径和高度表示；

片炭：规则片状的竹炭，用长度和宽度表示；

碎炭：破碎的形状不规则的竹炭；

颗粒炭：颗粒直径一般为 1.0~20.0mm 的竹炭；

粉末炭：颗粒直径一般小于等于 1.0 mm 的竹炭。

2.3.2 胶合分类法

原竹炭：以竹材为原料高温炭化得到

2.2.7 Bamboo charcoal

Bamboo charcoal is black solid porous product made of carbonized bamboo with limited oxygen conditions (Fig 2-3). The raw materials for bamboo charcoal manufacture are usually 4-year grown meso bamboo *(Phyllostachys edulis)* and its processing residues in China, traditional brick-clay kiln produced bamboo charcoal features with a high density and calorific value, but average bamboo charcoal still couldn't reach the quality standard of binchotan charcoal, meanwhile the specific surface area of bamboo charcoal can be upto 380 m^2/g. Bamboo charcoal are generally produced through drying, pre-carbonizing, carbonizing, and calcinging four stages by carbonizing bamboo.

Now many binchotan charcoal are produced from bamboos because a refinery processing has been employed to improve its performance successfully.

2.2.8 Bamboo charcoal briquette

Bamboo charcoal briquette is a compressed block of bamboo charcoal powder with/without other solid bio-fuels using or not using additives or adhesives (Fig 2-4). Each component in briquette serves an important purpose, giving briquette each of their distinctive qualities. Compressed block can be kinds of hollow cylinders that are easy-to-ignite.

Bamboo charcoal briquette can be subdivided into carbonized bamboo powder briquette (CBPB) and compressed bamboo charcoal powder briquette(BCPB), respectively, according to the production process.

Bamboo charcoal briquettes can be

的固体炭。

成型竹炭（A 型）：以竹粉为原料，经干燥、挤压成型再经炭化而制得的固体。

成型竹炭（B 型）：以竹炭粉为主要原料，添加胶黏剂混合，经成型干燥等工序而制得的固体。

2.3.3 工艺分类法

（1）按炭化温度分

低温竹炭：炭化温度≤ 600℃的竹炭。

中温竹炭：炭化温度 600~800 ℃的竹炭。

高温竹炭：炭化温度≥ 800℃的竹炭。

（2）按导电性分

导电竹炭：体积电阻率≤ $10^2\Omega \cdot m$ 的竹炭。

弱导电竹炭：体积电阻率介于 10^2~$10^7\Omega \cdot m$ 之间的竹炭。

非导电竹炭：体积电阻率≥ $10^7\Omega \cdot m$ 的竹炭。

2.3.4 用途分类法

日用保健竹炭：家居用品、洗涤护肤、食品保鲜、医疗用品等。

环境保护竹炭：空气净化、饮用水净化、污水处理等。

建筑装饰竹炭：调湿、工艺品、涂料、炭板等。

农林园艺竹炭：添加剂、土壤改良、无土栽培、植物生长促进剂等。

其他用途竹炭：煮饭、电磁屏蔽、燃料、活性炭、食品、电极等。

produced through processes: (a) Carbonization of bamboo powder briquettes (BCB-a, CBPB), and (b) Compression of bamboo charcoal powders (BCB-b, BCPB).

The raw material for process (a) can be compressed bamboo powder with or without other biomass powder blends and mixtures. The carbonized product works as a replacement to wood charcoal with extended combustion duration, high calorific value, smoke-free and explosive noise-proof, makes it perfect for industrial use.

The raw material for process (b) can be bamboo charcoal powder with or without other fuel blends and mixtures. The carbonized product are prone to underperform in combustion property but quite economically suitable for domestic applications.

2.2.9 Biochar

Biochar is a carbon-rich substance that is obtained by burning organic material from agricultural and forestry wastes (also called biomass) in a thermal decomposition (generally < 700℃) in an oxygen-limited environment. Although it looks a lot like common charcoal, biochar is produced using a specific process to reduce contamination and safely store carbon. During pyrolysis organic materials, such as wood chips, leaf litter or dead plants, are burned in a container with very little oxygen. As the materials burn, they release little to no contaminating fumes. During the pyrolysis process, the organic material is converted into biochar, a stable form of carbon that cannot easily escape into the atmosphere. The energy or heat created during pyrolysis can be captured and used as a form of clean energy.

In terms of physical attributes, biochar

图 2–5 粉末炭、颗粒炭、片炭、筒炭
(Fig 2-5 Powder, particular, flake and tubular bamboo charcoals, from top to bottom)

is black, highly porous, lightweight, fine-grained and has a large surface area. Applying biochar to degraded soils can help enhance its quality, for example, enhancing soil structure, increasing water retention and aggregation, and reducing greenhouse gas emissions, such as carbon dioxide (CO_2), nitrogen oxide (NO_x), and methane (CH_4).

2.3 Classification of bamboo charcoals

2.3.1 Classification according to shape and size (Fig 2-5)

Flake bamboo Charcoal is usually expressed according to length and width, and it's size is (width×length) of (30-50)m×(50-100)m bamboo charcoal flakes can also be provided in other specification upon request by supplier and purchaser.

Size distribution of granules and powders should be minimum 80% of which shall be fall in the range, i.e. minimum 80% of bamboo charcoal granules with the size of 2-4 mm can be regarded as Grade II, otherwise, re-grading or reprocessing is required to reach the corresponding grade.

Sometimes Bamboo charcoal rubble(also as broken bamboo charcoal) is also added to the classification, the rubble are irregular-shape with/without random size distribution.

Tubular bamboo charcoal is expressed according to length and diameter, and the length is usually greater than or equal to the diameter.

2.3.2 Classification according to processing

Bamboo charcoal can be classified as

2.4 竹炭与其他炭

2.4.1 竹炭和木炭

木炭由于天然林保护，其制造生产在我国有所限制，而竹炭由于是利用速生竹材烧制而成的，政府是鼓励的。竹炭和木炭具有以下区别：

首先，竹炭和木炭的原材料不同。竹材的种类相比木材少，目前主要以毛竹为主（也可采用其他竹种）烧制竹炭，从开展研究和应用推广意义上来讲，竹炭更具统一性。其次，两者的微观构造不同。在电子显微镜下观察，微孔排列、分布和孔径大小、孔容都不同。然后，基本的理化性能和应用领域有差异。如：一般来说木炭（白炭）的含碳量略高于竹炭，燃烧热值高，所以更适宜用作燃料；而竹炭的灰分含量一般比木炭高。

2.4.2 竹炭与煤炭

从原材料方面看，构成竹炭和煤的主要元素基本相同。煤中碳（C）含量一般在60%~98%，氢（H）含量一般在0.8%~6.6%，氧（O）含量一般在0.72%~16.1%，氮（N）含量一般在0.5%~1.5%，硫（S）含量一般在0.28%~0.58%。根据本书作者研究和查阅文献得出竹炭中碳含量一般在70%~93%，氢含量一般在0.3%~3%，氧含量也一般在2.30%~46.01%，氮含量一般低于1%，硫含量一般低于0.2%。

竹炭的热值能基本能与质量最好的煤相媲美。碳是决定发热量高低的主要元素，含碳量升高，固定碳含量增大；氢元素对

primitive bamboo charcoal and bamboo charcoal briquette according to processing.

Primitive bamboo charcoal: as-carbonized bamboo charcoal with no extra processing, generally referred as bamboo charcoal.

Bamboo charcoal briquette can be subdivided into carbonized bamboo powder briquette (CBPB) and compressed bamboo charcoal powder briquette (BCPB), respectively, according to the production process.

2.3.3 Classification according to manufacturing process

Based on the carbonization temperature, it can be classified as low (≤600 ℃), medium (600-800 ℃), and high (≥800 ℃) temperature carbonized bamboo charcoal; and accordingly it can be classified as conductive ($\leqslant 10^2 \Omega \cdot m$), semiconductive ($10^2$-$10^7 \Omega \cdot m$) and non-conductive ($\geqslant 10^7 \Omega \cdot m$) bamboo charcoal. There is a positive correlation between the conductivity of bamboo charcoal and carbonization temperature of bamboo materials.

2.3.4 Classification according to applications

Bamboo charcoal is a versatile biochar for its multiple applications. It can be used as solid fuel, and also served in daily healthcare(i.e. household products, food storage, personal care products, medicine products), environmental protection (i.e. air purification, drinking water purification, waste water treatment), architectural decoration (i.e. humidity conditioning,painting and coating, bamboo charcoal-based panel, handicrafts), gardening and farming (i.e. soil improvement or amendment, soil-less cultivation), and

挥发分含量影响较大，其含量升高，挥发分含量增大，同时也影响发热量；氧元素对热值及其他工业分析项目影响不大。

竹炭比煤更清洁、环保。竹炭的硫元素含量普遍较煤低，硫元素主要存在于灰分和固定碳之中，其含量与固定碳和灰分有关。

（1）竹炭粉与煤炭粉

竹炭粉是由竹炭经过加工而制得的，煤炭粉是煤经过加工而制得的，两者可以用扫描电镜（SEM）来观测判定区别。在显微镜下观察煤炭粉结构未见有孔隙结构，而竹炭粉有不规则特殊的六边形孔隙结构。

（2）竹炭的热值与煤炭的热值

影响竹炭和煤热值因素有很多，其中水分、碳含量、灰分和硫分对煤炭发热量的测定影响较大。竹炭和煤的种类不同其热值也不同，根据炭化程度煤可分为褐煤、烟煤和无烟煤，热值介于 12~34MJ/kg 之间。通过对 30 余种不同竹种的竹炭热值测试，得出其热值介于 26~33MJ/kg 之间。

2.4.3 竹炭和备长炭

竹炭最早应该是诞生在中国，南宋大诗人陆游《老学庵笔记》卷一中有记载："北方多石炭，南方多木炭，而蜀又有竹炭，烧巨竹为之，易然无烟耐久，亦奇物。"竹炭拥有密度大、热值高等特点，但类似备长炭的性能，市场上一般的竹炭的硬度、密度和固定碳含量都达不到"备长炭"的指标。因此，市场有"竹备长炭"或者"备长竹炭"的叫法不合理。

备长炭为白炭的一种（一种白炭的代名词），所使用的炭材为乌刚栎（中

other areas (i.e. electrode, electromagnetic shielding). More detailed applications can be available in Chapter 5.

2.4 Bamboo charcoal and other charcoal products

2.4.1 Bamboo charcoal and wood charcoal

Wood charcoal industrial is facing raw material shortages because of the forestry preservation and environmental protection. Fortunately, owing to fast growth of bamboo, it is encouraged to develop this bamboo-based product: bamboo charcoal, the profitable business in the technological fast lane has them charging full speed ahead in China mainland. The bamboo species and materials are relatively limited as compared to vast number of woods, which makes bamboo charcoal quite consistent in quality and performance in scientific research and massive production and promotion.

According to electron microscope observation, micro-pores in bamboo charcoal differs in pore array, size and volume distribution, which is essentially differentiated from wood charcoal, furthermore, their fundamental properties and applications are different as well. In many cases, wood charcoal(white charcoal) of higher fixed carbon characterized and better calorific value, is an appropriate fuel for energy supply, at this point bamboo charcoal of high ash content is not a perfect candidate.

2.4.2 Bamboo charcoal and coal

From the point of chemical composition, there are no obvious difference between bamboo charcoal and coal, but the contents are really

国）、姥芽栎（日本），炭质硬、组织细密且坚硬、比重大，1.2g/cm^3 左右，固定碳含量 93.0%~96.0%，一般竹炭达不到这几项指标。

竹炭的最高炭化温度一般低于 1000℃，而备长炭的炭化温度一般都在 1000℃以上；竹材的灰分含量本身高于木材的灰分，炭化过程中仍保留在炭中。此外，竹炭的比表面积可高达 380m^2/g，高于备长炭。

2.4.4 竹炭与竹质活性炭

竹质活性炭（Bamboo-based activated carbon）是竹材原料经炭化、活化等工艺制成的活性炭，和竹炭相比，两者主要有以下区别：

（1）工艺

竹炭只经过炭化阶段，而活性炭要经过炭化、活化两个阶段。

（2）孔隙结构

竹炭内部大中小孔分导管、维管束、薄壁组织侧壁上的小孔，竹炭的孔隙以大孔、中孔为主，其直径以 200nm 左右为主；而竹质活性炭以微孔占主导地位，按孔隙直径大小可分为三类：大孔（≥ 50nm），约占总孔容积的 10%~30%，微孔（≤ 2nm）约占总孔容积的 60%~90%，中孔又称过渡孔（2~50nm），约占总孔容积的 5%~7%，孔隙平均直径约为 1.5nm。

（3）比表面积

一般竹炭比表面积在 50~400m^2/g 之间；而用物理或化学方法进行二次活化制成竹炭活性炭，比表面积可达到 500m^2/g 以上。竹炭的比表面积比竹质活性炭小。

（4）产品性能

far from each other. As to coal, content of carbon (C), Hydrogen (H), nitrogen (N), and sulphur (S) is approx. 60% - 98%, 0.8% - 6.6%, 0.5% - 1.5%, and 0.28% - 0.58%, respectively. Referring to bamboo charcoal, content of carbon (C), Hydrogen (H), nitrogen (N), and sulphur (S) is approx. 70% - 93%, 0.3% - 3.0%, 0 - 1.0%, and 0 - 0.2%, respectively, by which it can be concluded that bamboo charcoal is cleaner and more environment friendly than coal because of its extremely low content of sulphur and nitrogen. While oxygen (O) content, both of bamboo charcoal and coal, varies significantly from a few to dozens of percentages.

Carbon (C) content is the main contributor to fixed carbon content, which is critical to calorific value; hydrogen (H) content is the main contributor to volatile matter content, affecting calorific value as well, but the oxygen (O) content exerts little impact to the value. Sulphur (S) can be found in existence from fixed carbon and ash.

Electron microscope observation can be introduced to distinguish from one another. Hexagonally structure pores in random arrangement can be seen from the microscopic pattern of bamboo charcoal, while no similar result can be found in coal.

As mentioned, calorific value is mostly affected by moisture, fixed carbon, ash and sulphur content. The calorific value of coal: lignite, bituminous or anthracite, measured to fall in a wide range: 12-34 MJ/kg, and that of bamboo charcoal is far narrow in contrast: 26-33 MJ/kg, a statistical analysis of over 30 bamboo charcoals.

2.4.3 Bamboo charcoal and binchotan charcoal

Bamboo charcoal may well originate in China, eminent poet *Lu You* (1125 - 1210 AD)

竹质活性炭的亚甲基蓝脱色和碘的指标远高于竹炭，但竹炭的密度、硬度较高且不易碎。

（5）产品类型

竹炭有筒炭、片炭、碎炭、颗粒炭、炭粉等；而竹质活性炭一般只有粉末状（直径≤ 1mm）、颗粒状（直径≤ 8mm）和柱状等。

（6）吸附对象

竹炭的孔隙以大孔、中孔为主，而竹质活性炭以微孔占主导地位。因此，竹炭可吸附的物质种类则多于竹质活性炭。由于许多微生物可进入竹炭微孔，因此竹炭常作为废水处理微生物的培养载体，也可用于饮用水的重金属吸附。而竹质活性炭以微孔为主，主要用于吸附室内甲醛等有害物质，其吸附速度更快、效果更明显。

of the Southern Song Dynasty (1127 - 1279 AD), wrote in his masterwork *Jottings from the hut of everlasting learning (laoxuean biji,* pubulished afterlife by his son in 1228 AD*):* *"Stone charcoals (namely coals) are common in northern China, wood charcoals are common in southern, and bamboo charcoals are common in Sichuan (a western province in China). Bamboo charcoal, obtained by burning of giant bamboo (maybe Dendrocalamus sinicus), is easy to ignite, smoke-free and durable in combustion. Quite a marvel."*

Bamboo charcoal of high density and calorific value shows similar property that of binchotan charcoal, nevertheless, average bamboo charcoal available in market falls short in hardness, density, and fixed carbon content.

Binchotan charcoal is one kind of white charcoal using oak *(Quercus phillyraeoides)* as raw material for carbonization. The charcoal show great hardness (as hard as cast iron), high density (1.8 - 1.85g/cm^3) and high fixed carbon content (93% - 96%) that difficulty to be attained by bamboo charcoal, however, specific surface area of bamboo charcoal can reach up to 380 m^2/g, is virtually higher than that of binchotan charcoal, whose carbonization temperature is greater than 1 000℃, while it generally is lower than 1 000℃ for bamboo charcoal manufacturing.

2.4.4 Bamboo charcoal and bamboo based activated carbon

Bamboo charcoal and bamboo based activated carbon differs in multiple facets:

(1) Manufacturing process

Bamboo charcoal is prepared by one-step carbonization procedure, meanwhile two-step procedure (carbonization and activation) is typical method for bamboo based activated carbon manufacturing.

(2) Pore structure

The pores of as-obtained bamboo charcoal are almost composed of macroproe (diameter ≥ 50 nm), derived from vessels and vascular bundles of bamboo, and mesopore (diameter ≥ 2 nm, ≤ 50 nm), derived from tiny holes in thin cell wall of parenchyma. The average diameter is approx. 200 nm.

After activation processing to bamboo charcoal, the pore size distribution completely changed in bamboo based activated carbon: macroproe (diameter ≥ 50 nm, 10 % - 30 % of total pore volume), mesopore (diameter ≥ 2 nm, ≥ 50 nm, 5 % - 7 % of total pore volume), micropore (diameter ≤ 2 nm, 60 % - 90 % of total pore volume), the average diameter is approx. 1.5 nm.

(3) specific surface area

Specific surface area of ordinary bamboo charcoal ranges from 50 m^2/g to 400 m^2/g, once activated, either physically or chemically enhancement, the area (of bamboo based activated carbon) can reach to 500 m^2/g or higher, much higher than that of bamboo charcoal.

(4) Properties

It mainly refers to the adsorption of organic compounds, for example, the adsorption capacity of iodine or methylene blue of bamboo based activated carbon outperforms that of common bamboo charcoal. As far as density or hardness concerned, bamboo charcoal in turn dominate the results.

(5) Product format

Bamboo charcoal can be classified as tubular, flake, granule, and powder product form, standard practice for bamboo based activated carbon is powder (diameter less than 0.18 mm), granule (diameter less than 8 mm), or briquette form.

(6) Adsorptive

Macropore and mesopore are typical structure units that constitute bamboo charcoal, while it is micropore for bamboo based activated carbon, which contribute to their huge difference in adsorption performance.

Bamboo charcoal can absorb various kinds of hazardous organic compounds with different molecular sizes, even microbes are easy to enter the pores, put another way, it can be used as microbe-carrier, which is extremely helpful in bacteria culture for biological approach to waster water management. Bamboo charcoal are at times applied to drinking water purification: adsorption of heavy metals.

On the other side, micro-pore dominated bamboo based activated carbon are chiefly used to absorb smaller molecules in a far more effective and efficient way, for instance, eliminating toxic formaldehyde from indoor air.

3 竹炭生产

Production of Bamboo Charcoal

3.1 竹炭原料与得率

3.1.1 原 料

竹炭原料一般选用4年生（含）以上的毛竹材及加工剩余料（竹根、竹枝、竹梢、竹条、竹屑），也可选用其他竹种来生产竹炭。其中竹材的弯曲、变形、裂纹、虫害、虫眼、霉变等缺陷都不影响竹炭的生产，但有特殊用途的竹炭，对原材料仍有要求。

3.1.2 得 率

（1）得率

竹炭的得率有气干得率和绝干得率两种表示方法，在实际生产中通常采用气干得率来表示，气干得率小于绝干得率。气干得率是指绝干重量的竹炭占气干竹材重量的百分比。其计算公式如下：

$$Y_1= m_0/m_1 \times 100\% \qquad （式 3-1）$$

式中：

Y_1——表示竹炭气干得率，%；

m_0——表示绝干竹炭的重量，g；

m_1——表示气干竹材的重量，g。

绝干得率是指绝干重量的竹炭占绝干竹材重量的百分比。其计算公式如下：

$$Y_2= m_0/m_2 \times 100\% \qquad （式 3-2）$$

式中：

Y_2——表示竹炭绝干得率，%；

m_0——表示绝干竹炭的重量，g；

m_2——表示绝干竹材的重量，g。

（2）影响因素

竹炭的得率同炭化设备、炭化工艺

3.1 Raw material and yield for production

3.1.1 Bamboo material

Commercial cultivation of meso bamboo (Phyllostachys edulis) is the earliest and most developed in China, taking the lead in planting area and yield every year (70 % of China's bamboo planting area). 4-year meso bamboo (and its processing waste) is the optimal choice for bamboo carbonization according to production experience and research findings.

There is no particular requirements for bamboo material, whatever defect or deficiency, e.g. wrapped, twisted, deformed, cracked, pest or mildew damaged, it is all undoubtedly suitable for carbonization to manufacture bamboo charcoal. Yet, for all that, some limits may be proposed upon bamboo material quality that producing for special purpose.

3.1.2 Yield

(1) Yield definition

Yield for production of bamboo carbonization, is commonly expressed as the percentage, by oven-dry weight, of bamboo charcoals obtained from the original bamboo weight. There are two expressions for yield: air-dried yield (Y_1), and oven-dried yield (Y_2), and the former is commonly adapted in practical production, which is also smaller than the later.

The air-dried yield (Y_1) can be calculated in compliance with formula (3-1):

$$Y_1 = m_0/m_1 \times 100\ \% \qquad (3\text{-}1)$$

Where,

Y_1 refers to air-dried yield expressed in percentages;

（炭化温度、炭化速度、保温时间）、竹材自身条件（竹种、竹龄、竹材的部位）等因素有关。

竹炭的得率与炭化设备有关，一般砖土窑比机械窑生产的竹炭密度要大，所以得率也相应高。

竹炭的得率也与炭化工艺有关，尤其是炭化温度最为显著，一般随着炭化温度的升高而呈下降的趋势，也随着炭化速率加快而下降。在同一炭化温度和速率条件下，随着保温时间延长，竹炭的得率也随之降低。

竹炭的得率与竹种有关但没有明显规律，一般随着竹龄的增加而增加，在同一炭化温度条件下，从竹材的基部到梢部，竹炭的得率略呈增大趋势。

3.2 竹炭设备与类型

竹炭的生产设备主要有砖土窑和机械窑两大类型。各种砖土窑和机械窑又有立式和卧式之分。

代表性的竹炭设备主要分布在中国和日本。

3.2.1 中 国

（1）砖土窑

砖土窑（图 3–1）是 20 世纪 90 年代末和 21 世纪初中国的竹炭生产设备，主要以泥土、砖和石灰等为材料砌成的，以立式为多。在结构上，有单排连窑和双排连窑背靠背两种形式，通常依据地形而定。窑体外形呈柱状，窑底为近似圆形，窑顶为拱形，窑门向外延伸，窑体的后部有烟囱口；从窑门往里依次是燃烧室、炭化室、烟囱三部分，燃

m_0 refers to oven-dried weight of bamboo charcoals expressed in grams;

m_1 refers to air-dried weight of bamboo materials expressed in grams.

The oven-dried yield (Y_2) can be calculated in compliance with formula (3-2):

$$Y_2 = m_0/m_2 \times 100\% \quad (3\text{-}2)$$

Where,

Y_2 refers to oven-dried yield expressed in percentages;

m_0 refers to oven-dried weight of bamboo charcoals expressed in grams;

m_2 refers to oven-dried weight of bamboo materials expressed in grams.

(2) Factors effecting yield

Yield of bamboo charcoal is often influenced by carbonization equipment, parameters (temperature, time) and bamboo materials (species, age, part). The yield decreases as the temperature increases, that is, rapid decrease in yield can be seen below 500℃ carbonization, and the fall is slow and steady above 500℃ . The yield also decreases with the quick carbonization rate. Under same carbonization temperature, the yield decreases as the carbonization time extended.

Yield of bamboo charcoal carbonized in brick-earth kiln is higher than that in automatic mechanical kiln, so does the density of bamboo charcoal. The bamboo species has effect, but not apparently, on yield of bamboo charcoal, but as bamboo grows (age increase), the corresponding yield increases. Under same carbonization temperature, the yield rises from bamboo base to tip.

3.2 Production equipment

Brick-earth kiln and mechanical kiln are widespread facilities in bamboo charcoal

烧室前侧通进风口和燃料投入口。

砖土窑制炭过程主要通过操作人员“眼观鼻嗅”来判断，难以控制竹炭生产工艺，导致产品质量参差不齐。也有少量砖土窑装有自动进出料的装置、温度显示控制仪、窑内热量及可燃气体回收循环系统，能比较准确地控制工艺，稳定产品质量。

①立式炭窑

a. 外形结构：窑墙高约 2m ；窑底为近似圆形，纵深 3.5~4.0m，最宽处约为 3.0m ；窑顶为拱形，拱高约 0.6m ；

b. 窑体容积：每窑 4~5t 竹材；

c. 窑体材料：砖、泥土和砂石等；

图 3–1　砖土窑

Fig 3-1　Brick-earth kiln for bamboo charcoal production

manufacturing, both kilns can be further classified as tower style and horizontal style, and the representative kilns are in China and Japan.

3.2.1　Equipment in the Chinese mainland

(1) Brick-earth kiln

Brick-earth kiln (Fig 3-1) is the leading facility used to produce bamboo charcoal at the turn of the 21st century, which is built by brick, soil and cement, etc. Tower style kiln is much popular at that time. Both single-rowed and back-to-back twin-rowed styles are often seen in field production. The tower-like cylinder-ed kilns are featured with near-round bottom base, arched top, outward door in front, and stack nozzle (of chimney) at the back. There are combustion chamber, carbonization chamber, and chimney from the outside in. Blast and fuel inlets are usually in front side of combustion chamber.

See and smell during production is quite helpful as rule of thumb used by experienced workers, but the results does not always turn out fail-safe, the quality in-conformity occurs on occasions. Automatic conveyer, temperature controller, and in-kiln heat and combustible gas recovery system sometimes are also quipped for the purpose of accurate control and quality consistency improvement.

① Tower style brick-earth kiln

Features of tower style brick-earth kiln:

Structure: kiln wall of 2.0 m in height, near-round bottom base with 3.5-4.0 m in length and 3.0 m at its widest, and arched kiln top of 0.6 m in height;

Capacity: 4-5 tons of bamboo materials each production batch (cycle);

d. 操作控制：主要通过操作人员“眼观鼻嗅”来判断，辅助温度显示控制仪；

e. 生产周期：筒炭、片炭生产周期20~25d，碎炭生产周期 7~10d；

f. 燃料消耗：用竹木材加工剩余物作为燃料，木竹煤气未回收利用；

g. 竹炭得率：15%~18%；

h. 竹炭质量：质量不均匀、差别大。

②卧式炭窑

a. 外形结构：窑体为长方体，窑墙高约 1~1.2m，长 2~3m，宽约 1.5m；窑顶为拱形，拱高约 0.6m；窑门形状为方形，设在窑体长度方向上，高和宽均约 0.5m；烟囱上出烟口位置在窑体后部，竹材和竹炭从窑顶部通过滑轮进出。

b. 窑体容积：每窑 1~1.5t 竹材，以竹片为主，成捆处理；

c. 窑体材料：耐火砖或普通砖、泥和砂石等材料；

d. 操作控制：装有温度显示控制仪显示和控制工艺；

e. 生产周期：生产周期 6~7 天；

f. 燃料消耗：用竹木材加工剩余物作为燃料，木竹煤气未回收利用；

g. 竹炭得率：20%~23%；

h. 竹炭质量：质量均匀。

（2）机械窑

生产竹炭的机械窑主要有立式和卧式两种类型。立式有单个和多个联排的。

①立式干馏釜

a. 外形结构：干馏釜由内外两个直立的金属圆筒组成，燃烧时产生的烟气在内外筒之间的夹层内流动，用以加热内圆筒中的竹材，使竹材炭化。

Building materials: brick, soil and cement;

Production control: *see and smell* method, auxiliary temperature probe;

production cycle: generally 20-25 days for tubular or flask bamboo charcoal, and 7-10 days for broken bamboo charcoal(rubble);

Energy consumption: bamboo/wood processing residues as fuel, combustible gases are not readily to recover;

Yield: 15%-18%;

Product quality: low quality consistency, variations in batches.

② Horizontal style brick-earth kiln

Features of horizontal style brick-earth kiln:

Structure: rectangular shaped kiln body: (1-1.2)m × (2-3)m × 1.5 m (height × length × width), arched kiln top: 0.6 m in height, kiln door: (2-3)m × 0.5m (height × length), and stack nozzle (of chimney) at the back of kiln. Bamboo materials are slid to load/unload with a roller on the kiln top;

Capacity: 1-1.5 tons of bamboo bundles each production batch (cycle);

Building materials: fireproof or regular brick, soil, sandstone and cement;

Production control: *see and smell* method, auxiliary temperature probe;

production cycle: generally 6-7 days;

Energy consumption: bamboo/wood processing residues as fuel, combustible gases are not readily to recover;

Yield: 20%-23%;

Product quality: improved quality consistency.

(2) Mechanical kiln

① Tower style kiln

Features of Tower style destructive distillation kiln

b. 窑体容积：每窑 0.7~0.8t 竹片或者竹筒；

c. 窑体材料：不锈钢材料；

d. 操作控制：装有温度显示控制仪显示和控制工艺；

e. 生产周期：1~2 天；

f. 燃料消耗：木（竹）煤气及余热回收利用；

g. 竹炭得率：18%~20%；

h. 竹炭质量：质量均匀。

②卧式炭化窑

卧式炭化窑有周期式和连续式 2 种。

第一种：周期式炭化窑（图 3–2）

a. 窑体结构：由一个横卧的金属圆筒组成，采用外热式加热，木（竹）煤气回收循环利用，竹醋液回收。

b. 窑体容积：每窑 0.4~0.5t 竹片；

c. 窑体材料：不锈钢材料；

d. 操作控制：自动控制炭化工艺；

e. 生产周期：1~2 天；

f. 燃料消耗：回收窑炉的烟气余热，提高能源利用率；

g. 竹炭得率：20%~22%；

h. 竹炭质量：质量均匀。

图 3–2　卧式周期式机械窑

(Fig 3-2　Periodic horizontal-style kiln)

Structure: the kiln is built by twin vertically-standing metal cylinders, the gaseous products pyrolyzed from bamboo can flow freely through the interlayer of the cylinders, which is used as heating source for carbonization;

Capacity: 0.7 - 0.8 tons of tubular or flask bamboo;

Building materials: stainless steel;

Production control: temperature probe and control system equipped;

Production cycle: generally 1-2 days;

Energy consumption: combustible gases and heat recovery;

Yield: 18%-20%;

Product quality: improved quality consistency.

② Horizontal style kiln

Features of periodic horizontal-style kiln (Fig 3-2):

Structure: a horizontal metal cylinder with wood/bamboo gases recovery system and bamboo vinegar collector, external heating;

Capacity: 0.4 - 0.5 tons of bamboo flakes;

Building materials: stainless steel;

Production control: temperature probe and control system equipped;

Production cycle: generally 1-2 days;

Energy consumption: combustible gases and heat recovery;

Yield: 20%-22%;

Product quality: improved quality consistency.

Features of Continuous style kiln (Fig 3-3):

Structure: the kiln built with combustion chamber, preheating chamber, and drying chamber. Automatic control system for heating control, proper temperature setting for chambers (-150℃ , 150-300℃ , 400-900℃ , respectively). As-pyrolyzed combustible gases

图 3–3 连续式炭化窑
Fig 3-3 Continuous -style kiln

第二种：连续式炭化窑（图 3–3）

a. 窑体结构：主要由干燥室、预热室和燃烧室组成，通过控制实现了各腔室的温度调节，可控制竹材炭化温度在 –150℃、150~300℃、400~900℃之间，分别实现竹材干燥、预炭化及炭化。外热式炭化窑将原料炭化产生的可燃气体进行充分热解，并把热量用于原料炭化、烘干和余热锅炉加热。炭化的温度可以通过混风器来控制，炭化的时间可以通过调整转炉的角度和转速来控制，实现原料炭化的数字化控制，使炭的质量更加稳定，也降低了人工成本。

b. 窑体容积：每套设备年产能 4000–5000t；

c. 窑体材料：不锈钢材料；

d. 操作控制：自动控制炭化工艺；

e. 生产周期：12~24h；

f. 燃料消耗：回收窑炉的烟气余热，提高能源利用率；

g. 竹炭得率：13%~15%；

h. 竹炭质量：质量均匀。

③砖土窑与机械窑

砖土窑和机械窑的具体比较见表 3–1：

recovered as energy source for carbonizing and drying inside the external heated kiln, and heat-recovery boiler. Mix-air controlled temperature, together with rotation angle/speed controlled carbonizing time, used for charcoal quality improvement and cost efficiency;

Capacity: 4000—5000 tons of bamboo materials, annual;

Building materials: stainless steel;

Production control: temperature probe and control system equipped;

production cycle: generally 12-24 hours per day;

Energy consumption: combustible gases and heat recovery;

Yield: 13%-15%;

Product quality: improved quality consistency.

③ Characteristics of brick-earth style kiln and mechanical style kiln

Characteristics of brick-earth style kiln and mechanical style kiln can be seen in Table 3-1.

3.2.2 Equipment in Japan

(1) Brick kiln

In Japan, quite a number of producer manufacture charcoals using brick kiln built

表 3–1 砖土窑和机械窑的特点

Table 3-1 Characteristics of kilns for bamboo carbonization

项目 Item	砖土窑 brick-earth kiln	机械窑 mechanical kiln
炭化形式 Heating form	内热式炭化 inner heating	外热式炭化 external heating
窑体温度 Temperature	介于 600~1000℃之间 600-1000℃	可以自主设定，不超过 1200℃ Programmable, technically ≤ 1 200℃
竹炭密度 Charcoal Density	较大 Larger	较小 smaller
窑内温差 In-kiln temperature difference	大、不均匀 Widely distributed	小、比较均匀 Narrowly distributed
投资成本 Investment scale	小 Relatively low	大 Relatively high
升温速度 Heating rate	升温慢，中间反应易控制 Slow, intermediate reaction easy to control	升温快，中间反应难控制 Fast, intermediate reaction difficult to control
生产效率 Productivity	低 Low	高 High
生产周期 production cycle	长 Long	短 Short
产品质量 Product quality	不均匀 Low quality consistency	均匀 Improved quality consistency
污染问题 Pollution	木（竹）煤气没有回收、有污染 Relatively harmful, emission of liquid and gaseous products	木（竹）煤气、竹醋液回收、环保 Eco-friendly, recycling gaseous products and bamboo vinegar

3.2.2 日　本

（1）砖土窑

在日本，有一定规模的厂家烧制竹炭多采用砖和泥砌制的砖窑，为提高效率和节能可数窑连在一起。较普及的是日本农林水产省林业试验场的砖土窑，带有竹醋液采集装置和消烟装置（图 3–4）。较有影响的有日本福井县小滨竹炭生产组合厂的砖炭窑及琦玉县日高市的炭窑等。

from bricks and soils, and multiple kilns may be built in tandem to improve productivity. Widespread Angle-formed brick kiln designed by Forestry Research Center of the Ministry of Agriculture, is equipped with bamboo vinegar collector and smoke eliminator (Fig 3-4). The most typical, influential kilns are located in Fukui, and Saitama, Japan.

(2) Eearth kiln - Satsuma kiln

Earth kiln - Satsuma kiln was developed by Forestry Research Center of Kagoshima, Japan, which uses volcanic rock and ash as

图 3-4 日本砖土窑
Fig 3-4 Brick-soil kiln in Japan

（2）泥窑—萨摩窑

日本鹿儿岛林业试验场研制的泥窑—萨摩窑是利用火山喷出物为材料堆积制造而成，其高 2m，长 4~5m，宽 3~4m。据文献资料介绍该泥窑造价低，但泥窑挖掘、建造工作量大，且窑身常会崩塌，所以须经常维护。

（3）车载式炭化炉

日本的车载式炭化炉，炭化室材料为钢材，外围采用耐火材料。烧制竹炭时，燃烧室产生的高温煤气进入炭化室，加热竹材并回收竹醋液，温度通常为 350~400℃。等竹醋液收集完成，燃烧室停止燃烧，在炭化室内导入少量空气，使竹材自燃，温度上升，进行精炼炭化。日本农林水产省林业试验场在欧美国家同类炉基础上改良研制出林试式移动炭化炉，炉壁材料为不锈钢，可拆卸，易搬动，呈圆锥形，由上中下 3 段组成。具体技术参数如表 3-2。

（4）干馏和炭化分开设备

北九州市森林生产联合体采用了一种干馏和炭化分开进行的装置，干馏在“竹醋液回收炉”中进行，该炉的外壳用钢板

building materials, with a rough dimension of 2 m × (4-5)m × (3-4)m (height × length × width). The affordable construction demands a good deal of workload of diggings to finish an earth kiln, as well as routine maintenance because of its unpredictable collapse at times.

(3) Movable style kiln

Movable style kiln is constructed by steel, and refractory (fireproof) materials. Coal-gas produced in combustion chamber will be introduced into carbonization chamber for bamboo heating, later after bamboo vinegar collected at temperature 350-400℃, ignition ceased and some air fed to induce spontaneous ignition of bamboo materials, at this point it moves on to refining stage of bamboo charcoal production. The movable style kiln was developed with reference to the groundwork of Western counterpart. The kiln wall is made of steel, easy-to-disassemble and movable, consisted of three cone-shaped structural units. Detailed parameters are listed in table 3-2:

(4) Independent style Destructive distillation - Carbonization kiln

An independent style Destructive distillation - Carbonization kiln was run by Forestry Production Complex of Kitakyushu, Japan. Destructive distillation occurs in

表 3–2 林试式移动炭化炉参数
Table 3-2 Parameters of the movable style kiln

项目 Item	1900 型 type1900	1200 型 type1200
底面直径 /mm Bottom diameter, mm	1900	1200
高 /mm Height, mm	1860	1800
重量 /kg Weight, kg	180	120
装料量 /kg Loading capacity, kg	1500	450
使用寿命 /a Service life, a	3~5	3~5

制成，内衬耐酸的氧化铝，容积 1.8m^3，可装 600kg 竹材，采用直接加热方式，温度控制在 200℃左右。该装置自动控制升温速度、最终温度和加热时间，通常干馏时间为 4h，可得竹醋液 240L；干馏结束后，炉内竹材用叉式升降机移入炭化炉，该炭化炉的内壁浇灌耐火混凝土，温控装置将炉内温度控制在 850~1000℃，燃烧时间 3h，由点火喷嘴点火使其自燃，每炉可得竹炭 120kg（竹炭得率约 20%）。

3.3 竹炭生产与加工

3.3.1 生产工艺流程（图 3–5）

3.3.2 竹材预干燥

竹材的含水率直接影响到竹炭产品的得率。竹材预干燥通常有气干和窑干两种方法。竹炭生产企业常采用气干的方法，成本低，但含水率比较难控制。据作者课

bamboo vinegar collecting chamber. The Kiln is built by steel, lined with acid proof alumina coating layer, with a volume of 1.8 m^3 and bamboo loading capacity of 600 kgs, programmable heating, and automatic control of temperature and time. Direct heating method is used to hold the temperature at 200℃ . It can collect 240 Liter of bamboo vinegar in a 4-hour destructive distillation period, after that, a follow-up transferring of bamboo to carbonization chamber, the chamber wall is cast by fireproof cements, adopting similar control system in bamboo vinegar collecting, whilst the temperature set at 850-1 000℃ , time at 3 hours, and igniting via a ignition burner. Typical yield is approx. 20 %, or 120 kg of bamboo charcoal each time, instead.

3.3 Production and Processing

3.3.1 Process and procedure (Fig 3-5)

3.3.2 Bamboo material pre-drying

Moisture content of bamboo materials

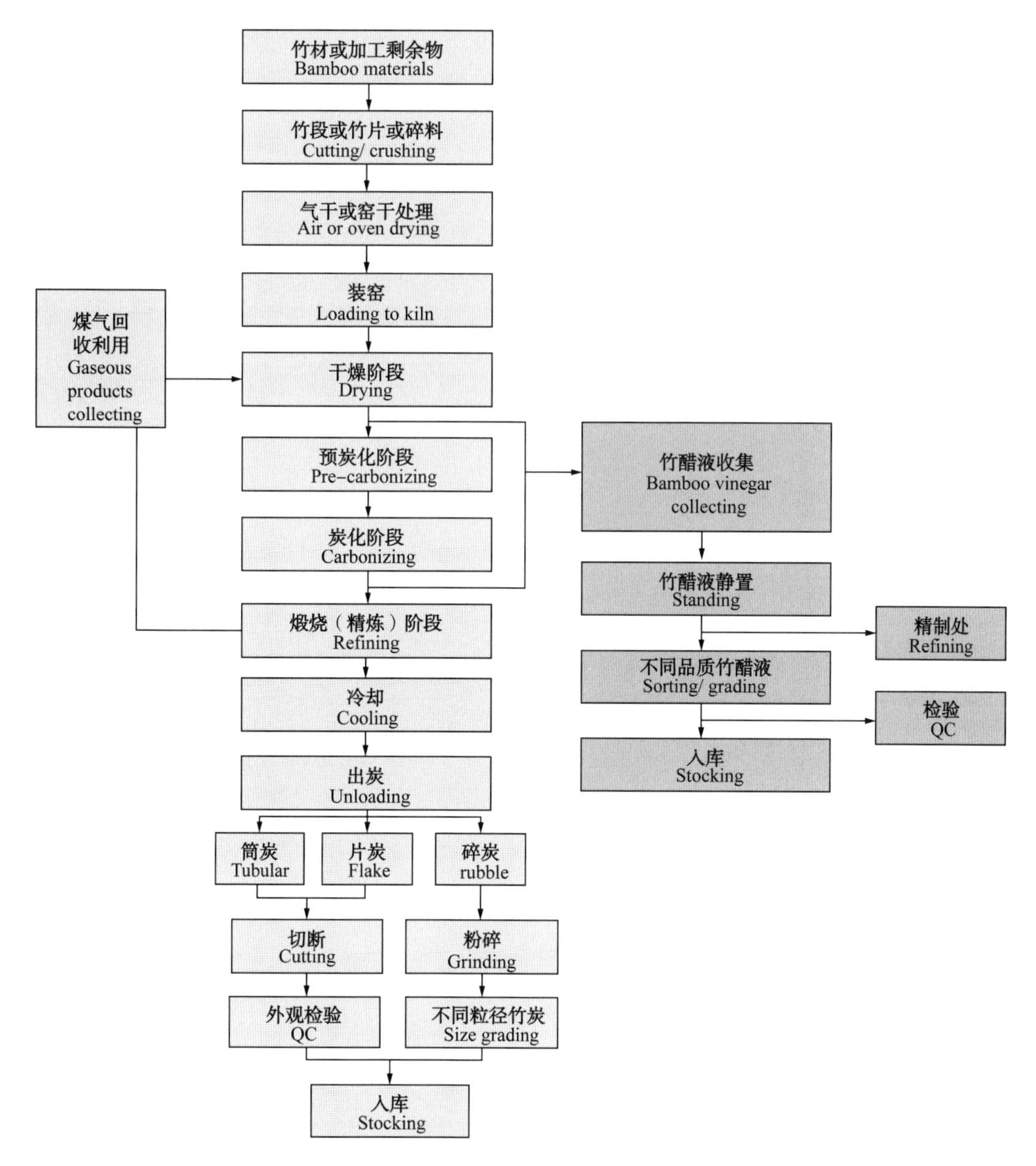

图 3-5 竹炭、竹醋液生产工艺流程

Fig 3-5 Typical process for bamboo charcoal/vinegar production

题组的研究表明，竹材的含水率控制在20%~25%之间，制得的竹炭综合性能指标较好。

3.3.3 竹材热解

竹材的热解是一个极其复杂的物理和化学变化过程，根据温度变化情况，大体

can exert a great influence over yield of the carbonized, it should be pre-dried to drive off water prior to carbonization by oven drying or air drying, and the latter is frequently applied due to better cost efficiency, on the other side of the coin, it is time consuming and mositure control tends to be difficult. Considering the authors' research and practical production,

可分为干燥、预炭化、炭化、煅烧四个主要阶段，其主要反应和产物见表 3-3，但四个阶段，没有明显的界线。

（1）竹材的干燥阶段

加热温度低于 150℃时，竹材分解的速度缓慢，主要是竹材组织中的吸着水受热蒸发逸散，竹材的化学组成没有明显变化；但随着加热时间的延长，竹材中的戊聚糖含量可能降低，竹材的物理力学性质有所改变。

（2）竹材的预炭化阶段

加热温度为 150~350℃，竹材受热分解速度加快，其中的化学组成发生明显的

moisture content of bamboo can be fixed at 20%-25% for overall consideration.

3.3.3 Bamboo pyrolysis (carbonization)

Pyrolysis is the thermal decomposition of materials at elevated temperatures in an inert atmosphere. It involves a series of physical and chemical changes. The Pyrolysis (carbonization) of bamboo can be divided into four stages: drying, pre-carbonizing, carbonizing, and calcining (refining), between which according to research and production, no clear-cut distinction can be drawn, the main reactions and products of the pyrolysis can be seen in Table 3-3:

表 3–3 竹材热解的四个阶段

Table 3-3 Stages of bamboo pyrolysis

项目 Item	干燥阶段 Drying	预炭化阶段 Pre-carbonizing	炭化阶段 Carbonizing	煅烧阶段 Refining
反应温度 Reacting temperature	<150℃	150~280℃	280~450℃	>450℃
反应物 Reacting substance (reactant)	自由水蒸发 Free water (evaporation)	半纤维素和纤维素 Cellulose and hemicellulose	纤维素和木素 Cellulose and lignin	热解残余物 Pyrolysis residues
液体产物 Liquid product	水分 Water	水、少量的乙酸、甲醇 Water, and small amount of acetic acid and methanol	大量的乙酸、甲醇、木焦油及其他有机物 acetic acid(large amount), methanol, wood tar, and complex organic compounds	很少 Rare
气体产物 Gaseous product	空气及少量的二氧化碳 Air, carbon dioxide(small amount)	二氧化碳以外，可燃性成分一氧化碳、甲烷 carbon dioxide and combustible gases (carbon monoxide, methane)	一氧化碳、甲烷、氢气等可燃性成分 combustible gases (carbon monoxide, methane, hydrogen)	很少 Rare
反应热 Enthalpy (thermal effect)	吸热 Endothermic (heat absorption)	吸热 Endothermic (heat absorption)	放热 Exothermic (heat release)	吸热 Endothermic (heat absorption)

分解反应，比较不稳定的组分如半纤维素受热分解生成二氧化碳、一氧化碳、水和少量的乙酸等物质。上述两个热分解阶段，需要外界供给热量，亦即竹材的吸热分解阶段。

（3）竹材的炭化阶段

加热温度升高至350~450℃，竹材热分解反应剧烈，伴随产生大量的热分解产物，生成的气体中，二氧化碳和一氧化碳的量逐渐减少，而碳氢化合物如甲烷、乙烯、烯烃类及各种活性高能的氢自由基和羟基，则逐渐增多；生成的液体主要有乙酸、甲醇、丙酮和竹焦油。这一阶段放出大量的反应热，称为放热反应阶段，而放热反应温度随加热速度而异。

（4）竹材的煅烧阶段

温度继续升高到450℃以上时，生成的液体产物已经很少，该阶段借助外部的热量对炭进行煅烧，并除去残留在炭中的挥发物质。

3.3.4 竹炭品质

竹炭品质主要同炭化设备、炭化工艺有关，其次同竹种、竹材立地条件、培育方式、部位等相关。开发竹炭必须适材适用、因地制宜等原则，根据其用途，选择工艺和对应的竹材。如竹炭咖啡棒、风铃工艺品宜采用小径级竹烧制。

3.3.5 贮存和运输

竹炭在存放过程中会吸附水分和有害物质，表面的孔隙在使用中会渐渐阻塞，会影响使用效果并带来二次污染，吸附过多水分还容易滋生细菌和霉菌。竹片炭和

Stage 1: Bamboo drying

At a temperature below 150℃, the thermal decomposition is slow, bamboo material itself is readily to dehydrate (evaporation of absorbed water in bamboo), there is no significant change in chemical compositions of bamboo, The content of pentosan (a leading component of hemicellulose) presumably drops for an extended period of carbonization, meanwhile, physical and mechanical properties changes accordingly.

Stage 2: Bamboo pre-carboniaztion

Bamboo pyrolysis accelerates considerably at a temperature range of 150-350℃, obvious thermal changes are revealed, the heat sensitive components (hemicellulose) are prone to degrade, forming CO_2, CO, H_2O, and a small amount of acetic acid also can be obtained in this stage.

Stage 3: Bamboo carbonizing

Pyrolysis of bamboo reacts intensely at the range of 350-450℃, accompanied by abundant decomposition products, that is, carbonaceous gases (CO_2, CO) are diminishing, hydrocarbon compounds (methane, ethene, and other alkene) and highly active radicals (hydrogen and hydroxyl) are emerging, liquid products (acetic acid, methanol, acetone and bamboo tar) are developing in this stage. The temperature of exothermic (heat release) carbonization varies depending on heating rate.

Stage 4: Bamboo calcining (refining)

Liquid products are rare when the pyrolysis temperature is beyond 450℃, the volatile matters in charcoal are eliminated and charcoal calcined (refined) through external heating in this stage.

3.3.4 Quality of bamboo charcoal

The quality of bamboo charcoal are

竹筒炭在太阳下曝晒容易开裂和破碎。

竹炭产品在运输过程中防止污损，不得受潮、雨淋，不允许与化学、腐蚀性物质混装。产品应放于干燥、洁净、通风的仓库里贮存，并远离火源；贮存时应按类别、规格、等级分别堆放，每堆应有相应的标记。

often affected by carbonization process, bamboo species, part, growth conditions and cultivation. Development of bamboo charcoal products should be in accordance with end applications, apply proper carbonization process based on bamboo material features and the place of origin.

Large-diametered meso bamboo (phyllostachys edulis) is mostly grown in the Chinese mainland (60 % of bamboo planting area), and undoubtedly the leading bamboo species for charcoal production, some small-diametered bamboo are also carbonized for particular purposes, such as coffee stirrer, and handicraft articles.

3.3.5 Storage and transport

The applicable national or regional safety guidelines for handling and storage shall be followed. During transportation, the packaged bamboo charcoal shall be handled with care to avoid moisture, sunlight, contaminant exposure or any damage to bamboo charcoal and its package.

Bamboo charcoal shall be stored with marking or labeling, in a dry, cool place (ambient temperature) away from sunlight, moisture and ignition conditions.

4　竹炭特性

Properties of Bamboo Charcoal

4.1 竹炭结构与组成

4.1.1 结构与形貌

竹炭结构是含石墨微晶的无定型碳结构，基本保持竹材的微观形态，有明显的塌陷现象，细胞壁组织上的微孔道是竹炭具有较大比表面积的主要原因，竹炭孔壁上为 20μm 左右的圆孔组织，其随着炭化温度的提高竹炭中的孔隙率增加。导管内壁存在类似层状石墨的结构；竹材基本组织的细胞间隙变小、消失，细胞壁变薄，细胞腔变大；薄壁细胞内壁变得光滑、干净。竹材的纤维鞘炭化后虽然也产生了部分孔隙，但仍然维持相对密实的状态，并且纤维细胞壁明显变薄。

竹炭横截面的 SEM 电镜图（图 4–1）可见维管束有不同程度的收缩，细胞间隙减小，维管束外鞘变得更致密，导管内沉积物减少，内表面光滑。维管束之间的基本薄壁组织细胞收缩变小，形成类似六边形的孔。纵截面上可见导管壁上的纹孔更明显，有大小不同的孔隙结构，主要以大孔、中孔为主。

4.1.2 元素构成

竹炭元素组成及含量同竹种、产地、竹材部位、炭化工艺等都有关。其元素组成主要有碳（C）、氢（H）、氧（O）、氮（N）、钾（K）、钙（Ca）、钠（Na）、镁（Mg）、铁（Fe）、铜（Cu）、锌（Zn）、铝（Al）、磷（P）、硼（B）、砷（As）、铅（Pb）、铬（Cr）、硅（Si）、钴（Co）、锰

4.1 Structure and constitutes

4.1.1 Structure and morphology

Bamboo Charcoal is a carbonecaeous material contains graphite crystallite, maintained a basic micromorphology of original bamboo, yet evident collapse of mirconstructure is still observed.

However, the structure of the bamboo was slightly changed in the process of carbonization, more specifically, cell walls thinned, cavity enlarged, intercell gap shrinked and even disappeared occasionally, parenchyma (basic unit of bamboo) smoothened, and inclusions reduced (pyzolyzed and vaporated). From the microscopic images of sectional surface of bamboo charcoal, the vascular bundles densed, Different pore sizes (mainly macropores and mesopores) are visually distinguishable, confirming hierarchical pore structure of the bamboo charcoal (Fig 4-1).

The cells and tissues of bamboo are carbonized to provide bamboo charcoal of abundant hexagonal micropores, increaseing specific surface area, which in turn enhances its absorption property.

4.1.2 Elementary composition

Main elements of bamboo charcoal include C, H, O, and N, the content varies according to the bamboo species. Content of trace elements in bamboo charcoal changes with carbonization process and equipment, and original bamboo materials, for example, the bamboo species, part, and place of origin. To date, there have been various elements

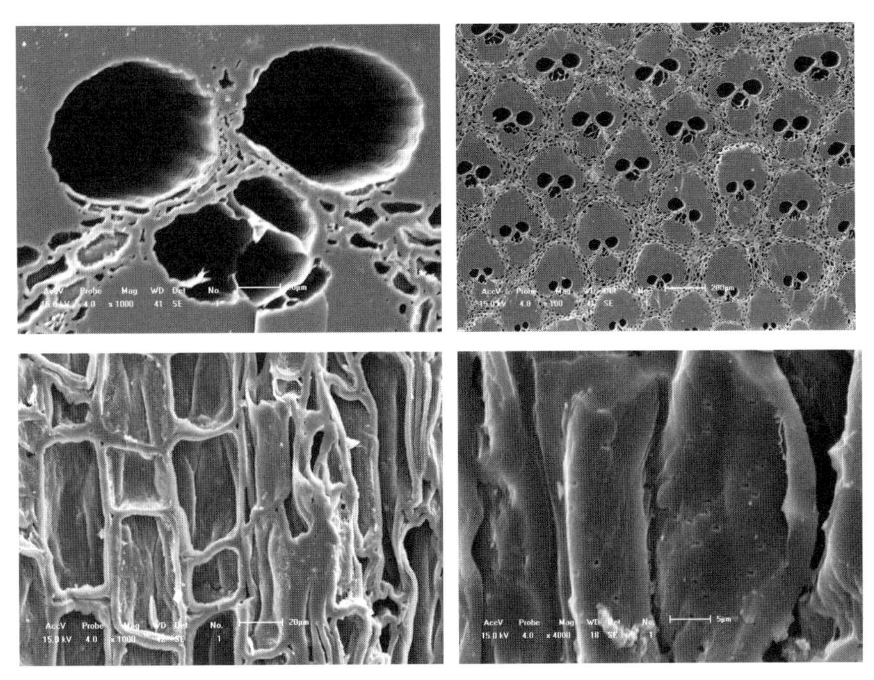

图 4-1 竹炭微观结构

Fig 4-1 Microstructure of bamboo charcoal

（Mn）、硒（Se）、硫（S）、锡（Sn）、镓（Ga）、钯（Ba）、镍（Ni）、锶（Sr）等。

4.2 竹炭的物理化学指标

4.2.1 含水率

（1）竹炭中的水分来源

竹炭中的水分通常有三种来源，一是在生产过程中吸水，二是在储存过程中吸收水蒸气，三是在生产、储运过程中非正常的吸水，如雨淋、雪盖等。

（2）定 义

含水率是指单位质量竹炭所含的水分质量，通常称为竹炭水分含量。在窑中竹炭含水率为 0，从窑体中刚取出的竹炭含水率约为 1%~2%，在空气中竹炭稳定的含水率为 8.0%~15.0%。

confirmed essential or relative to human health:

Group I, essential to human life: iron (Fe), iodine (I), zinc (Zn), selenium (Se), copper (Cu), molybdenum (Mo), chromium (Cr), and cobalt (Co);

Group II, maybe essential: nickel (Ni) and boron (B), manganese (Mn), silicon (Si) and vanadium (V);

Group III, possible toxic, and maybe necessary for certain functions: fluoride (F), lead (Pb), cadmium (Cd), mercury (Hg), arsenic (As), aluminium (Al), lithium (Li), tin (Sn).

4.2 Physical and chemical properties

4.2.1 Moisture content

(1) Definition

Moisture content is the amount of water present in a moist sample of a product like

（3）测定方法

通常用烘干法测定。即称取 1~5g（称准至 0.1mg）试样（粒径小于 1mm 或过 18 目筛），放入预先干燥至恒重的称量瓶中，试样在称量瓶底面厚度均匀。在 105℃ ± 5℃的电热恒温干燥箱中，干燥 3~4h，取出，放入干燥器中，冷却至室温（大约需 30min）后称量。然后进行检查性试验，每次干燥时间为 30min，直到试样质量恒定或质量变化小于 0.0050g 为止，在后一种情况下，必须采用前一次质量作为计算的依据。

（4）应用领域要求

竹炭含水率在许多应用领域是一个重要的指标，因此竹炭成品的干燥与密封是个不容忽视的环节。在不同的使用场合，对竹炭含水率的要求不同，作为烧烤燃料和吸附用竹炭要求水分越低越好，而作为土壤改良和地板调湿用竹炭等需要其能储存较高水分。

4.2.2 灰　分

（1）定　义

竹炭灰分是指煅烧至恒重后的残留物，是竹炭的无机组成部分，含量相对较多的有钾（K）、钙（Ca）、钠（Na）、镁（Mg）、硅（Si）和铁（Fe）、锌（Zn）、铬（Cr）等元素。

（2）影响因素

灰分含量高低主要受炭化温度影响，一般随着炭化温度升高而增大；还与竹材种类、不同部位等有关。通常 1.0%~6.5%。

（3）测定方法

将试样粉碎至 60 目以上，在 102~105℃的电热鼓风干燥箱中，干燥至恒重。将带

bamboo (charcoal), wood, soil or similar. Moisture content of bamboo charcoal that freshly produced in a kiln can be be regarded as 0 (accordingly, as oven-dry product), then reaches to 1%-2% when taken out of a kiln, and stabilizes at a level of 8%-15% in air.

(2) Moisture source

The porous bamboo charcoal are readily to absorb humid from air. The moisture (water) source inside bamboo charcoal can be attributed to absorption of moisture in production, storage and transportation, and improper exposures, such like rain or snow exposure.

(3) Measurement

The moisture content of bamboo charcoal is determined by removing moisture and then by measuring weight loss by oven-dry method.

Weight approx. 1-5 g of bamboo charcoal sample by a analytic scale(to 0.0001g), the sample size shall be <1mm or pre-screened by 18 mesh, and subsequently heated (105℃ ±5℃) in an oven to allow for the release of moisture for 3-4 h on end. Following this, the sample is cooled for approx. 30 min in the desiccator prior to reweighing. Moisture content is calculated by the difference in wet and dry weight. In this process, measuring accuracy and the resolution of the balance are extremely important. Careful consideration must also be given to maintain identical conditions, where temperature and duration are vital for generating precise and reproducible results.

(4) Requirements for applications:

Moisture is one of the key index for applications, which makes drying and sealing indispensable in the production. For example, it requests a lower moisture when bamboo charcoal is used as fuel or absorbent,

盖瓷坩埚在800℃下灼烧到恒重，称取试样1g（精确至0.0002g）放入灼烧后已知重量的坩埚中，并放入温度不超过300℃的高温电炉中，敞开坩埚盖，使炉温逐渐升到800℃，并在800℃±20℃的条件下灼烧2h，取出坩埚盖上坩埚盖，在空气中冷却约5min，放入干燥器中，冷却到室温再称量。以残留物质量占试样原质量的百分数作为灰分含量值。

（4）应用领域要求

作为活性炭的炭化料、燃料、成型炭燃料等方面需要灰分含量越低越好；而在饮用水净化、土壤改良等方面则需要灰分含量高一些。

4.2.3 挥发分

（1）定　义

竹炭在850℃高温下煅烧时释放的一氧化碳（CO）和甲烷（CH_4）等气态产物称为挥发分。

（2）影响因素

挥发分含量受炭化温度影响最显著，随着炭化温度的升高，其值呈现降低趋势，同时还与竹材种类、不同部位等有关。通常≤15.0%。

（3）测试方法

将试样粉碎至60目以上，在102~105℃的电热鼓风干燥箱中，干燥至恒重。将带盖瓷坩埚（测挥发分特制的坩埚高40mm，上口径为30mm，下口径为18mm，盖的外径为35mm，槽的外径29mm，外槽深4mm）在850℃下灼烧到恒重，称取试样1g（精确至0.0002g）放入灼烧后已知重量的坩埚中，盖上坩埚盖，放在坩埚架上送

conversely, product with higher moisture absorption capability is preferred to be applied in soil amendment, and moisture conditioning of flooring.

4.2.2 Ash content

(1) Definition

Ash Content is the inorganic residue remaining after the water and organic matter have been removed by heating, the remaining residue consists of oxides and salts containing anions such as phosphates, chlorides, sulfates, and other halides and cations such as sodium, potassium, calcium, magnesium, iron, and silicon from the bamboo charcoal.

(2) Factors affecting the ash content

It is mostly affected by carbonization temperature, technically it increases with the temperature rise, in addition, bamboo species and parts are also the factors that will influence the value. The ash content by the large goes to 1.0%-6.5%.

(3) Measurement

Bamboo charcoal shall be pulverized or ground to pass 60 mesh (250μm), and oven-dried prior to ignition. Weight approx. 1.0g of bamboo charcoal sample (to 0.0001g) for burning. Dry ashing procedures use a high temperature muffle furnace capable of maintaining temperatures of 800℃ or above. Water and other volatile materials are vaporized and organic substances are burned in the presence of the oxygen to give CO_2, H_2O and N_2. After 2 hour burning at 800℃±20℃ , most minerals are converted to oxides, sulfates, phosphates, chlorides or silicates. Following this, the sample is taken out the furnace to cool for approx. 5 min prior to transfer to the desiccator, reweigh the sample. The difference in weight before and after the

入温度预先加热到850℃的高温电炉中，立即严闭炉门同时开始计时，持续7min后取出，在空气中冷却5min后放入干燥器中冷却到室温再称量。所失去的质量占试样原质量的百分数作为挥发分（V）。

（4）应用领域要求

作为活性炭的炭化料、燃料用竹炭等需要较高的挥发分含量；屏蔽用竹炭需要低挥发分含量。

4.2.4 固定碳

（1）定 义

固定碳含量是指竹炭中有效碳元素的百分含量，一般以竹炭的绝干重量减去其灰分含量和挥发分含量计算得到。

（2）影响因素

固定碳含量受炭化温度影响最大，随着炭化温度升高而增大，同时还与竹材种类、不同部位等有关。通常固定碳含量≥75%。

（3）应用领域要求

固定碳含量是衡量竹炭质量的重要指标之一，其直接标志着竹炭的炭化程度。固定碳含量随着炭化终点温度升高而显著增加。当温度升高到一定程度以后，其固定碳含量增加趋缓，由此说明在烧制竹炭的过程中炭化温度的选择和控制对提高和稳定竹炭质量是十分重要的。一般领域应用都要求固定碳含量高。

4.2.5 密 度

（1）定 义

竹炭密度为单位体积的质量，有堆积密度、颗粒密度和真密度三种表示方法，

test is the percentage of the ash content.

(4) Requirements for applications

Low ash content is favorable for fuel and briquette application, raw material for activated carbon production, and high ash content for drinking water purification, soil amendment, etc.

4.2.3 Volatile matter content

(1) Definition

Volatile matter is essentially a measure of the non-water gases formed from a bamboo charcoal sample during heating. It is measured as the weight percent of gas (emissions) from a sample that is released during heating to 850 ℃ in an oxygen-free environment, except for moisture (which will evaporate as water vapor), at a standardized temperature.

(2) Factors affecting the volatile matter content

The volatile matter content is notably affected by carbonization temperature, often decreases as the temperature increases, in addition, bamboo species and parts are also the factors that will influence the value. The volatile matter content, as a rule, is no greater than 15.0 %.

(3) Measurement

Alike to ash content determination, same procedure can be taken to prepare the bamboo charcoal sample. Prior to testing, moisture content is driven off and recorded as the residual moisture content. Volatile matter testing occurs after the sample is dry. In the test, weighted sample is transferred to a muffle furnace, that is preheated to 850℃ , immediately subjected to high heat in an atmosphere of pure nitrogen gas for 7 min, and then bring out the sample, cool down for approx. 5 min prior to transfer to the

通常所指的密度是指真密度。实质体积不包括颗粒内部空隙体积和颗粒间的空隙体积。

堆积密度（ρ_b）又称填充密度，它是指在规定条件下，单位体积（以竹炭颗粒间的空隙体积、竹炭颗粒内部空隙体积和竹炭的实质体积之和为体积）的竹炭质量，即：

$$\rho_b = \frac{m}{V_b} = \frac{m}{V_{void} + V_{pore} + V_{true}} \quad （式 4–1）$$

式中：m——竹炭的堆积质量，g；

V_b——竹炭的堆积体积，cm^3；

V_{void}——竹炭颗粒间的空隙体积，cm^3；

V_{pore}——竹炭颗粒内部空隙体积，cm^3；

V_{true}——竹炭的实质体积，cm^3。

颗粒密度（ρ_P）是指在规定条件下，单位体积（以竹炭颗粒内部空隙体积和竹炭的实质体积之和为体积）的竹炭质量，即：

$$\rho_P = \frac{m}{V_{pore} + V_{true}} \quad （式 4–2）$$

真密度（ρ_T）是指规定条件下，单位体积（竹炭的实质体积，不包括颗粒空隙体积和颗粒间的空隙）的竹炭质量，即：

$$\rho_t = \frac{m}{V_{true}} \quad （式 4–3）$$

（2）影响因素

竹炭的密度大小同炭化设备、工艺（炭化温度、保温时间和炭化速度等）、竹材原料（种类、竹材部位和竹龄等）有关，通常为 0.2~1.2g/cm^3。一般地，传统的砖土窑生产的竹炭密度高于机械窑；炭化速度慢、保温时间长则相应的竹炭密度较大。

desiccator, reweigh the sample. The difference is the weight percentage lost as emissions during combustion, which should be the volatile matter in the sample.

(4) Requirements for applications

Bamboo charcoals of high volatile matter content is suitable for fuel applications and activated carbon manufacture, and the shielding-functioned products prefers a low volatile matter content.

4.2.4 Fixed carbon

(1) Definition

Fixed carbon is a measure of the amount of non-volatile carbon remaining in bamboo charcoal sample. It is generally calculated by other parameters (contents) determined in a proximate analysis, rather than through direct determination. Fixed carbon is the calculated percentage of material that was lost during the testing for moisture, volatile matter, and ash content.

(2) Factors affecting the fixed carbon content

Fixed carbon has the opposite trend of volatile matter with increasing carbonization temperature, because increases in the amount of volatile matter driven off of bamboo charcoal increase the relative amount of carbon, put another way, fixed carbon content increases with the carbonization temperature rise, it will ultimately reach to a plateau phase when the temperature is high enough, which signifies the optimal point to produce bamboo charcoal with better quality in actual production. Also, bamboo species and parts are the factors that will influence the content.

(3) Requirements for applications

Fixed carbon is one most important criteria for estimating the quality of bamboo charcoal,

（3）测定方法

对于不规则的片炭或者颗粒炭，可采用固体密度计（排酒精法）进行测定，得到竹炭真密度。对于竹炭粉可以测堆积密度，用量筒取已知质量炭粉，堆实后记下体积值，计算得到竹炭粉的密度。

（4）应用领域要求

通常竹炭的密度大，也就意味着强度高。如炭包、炭（床）垫、饮用水净化用竹炭等需要密度相对较大；而在土壤改良、炭基肥等领域则需要密度相对较低的竹炭。

4.2.6 酸碱性

（1）定　义

竹炭酸碱性指竹炭水煮溶液的酸碱度。非粉体的竹炭需经破碎后再水煮。一般测定竹炭的溶解 pH 值。

（2）影响因素

竹炭 pH 值一般同炭化工艺、不同部位等有关。通常 pH 值为 4.0~11.0。

（3）测试方法

称取未干燥的竹炭试样 5g（60 目以上），置于 100mL 的锥形瓶中，加入水 50mL，加热缓和煮沸 5min，补添蒸发的水，过滤，弃去初滤液 5mL。余液冷却到室温后用酸度计测定 pH 值。

（4）应用领域要求

酸性土壤改良用竹炭需要 pH 值高，反之，需要 pH 值低；在饮用水中和时需要 pH ≥ 7 的竹炭。吸附氨气等气体 pH 值低的竹炭更有效。

which, as a rule, is greater than 75.0 %. High fixed carbon of bamboo charcoal is required for most application areas.

4.2.5 Density

(1) Definition

Bulk density, also known as apparent density or volumetric density, is a property of powders, granules, or any other masses of particulate matter. It is defined as the ratio between the apparent volume and dry specimen mass of a sample (bamboo charcoal herein).

Bulk density (apparent density) of bamboo charcoal can be calculated by formula (4-1):

$$\rho_b = \frac{m}{V_b} = \frac{m}{V_{void} + V_{pore} + V_{true}} \qquad (4\text{-}1)$$

Where,

ρ_b refers to the bulk density of bamboo charcoal expressed by g/cm^3;

m refers to the mass of the bamboo charcoal expressed by grams;

V_b refers to the bulk volume of the bamboo charcoal expressed by cm^3;

V_{void} refers to the void volume between bamboo charcoal expressed by cm^3;

V_{pore} refers to the pore volume inside bamboo charcoal expressed by cm^3;

V_{ture} refers to the true volume of bamboo charcoal expressed by cm^3.

Particle density of bamboo charcoal is defined as the ratio between the unit volume and dry specimen mass of a sample, which includes the volume of the pores. It can be calculated by formula (4-2):

$$\rho_p = \frac{m}{V_{pore} + V_{true}} \qquad (4\text{-}2)$$

Where,

ρ_p refers to the particle density of bamboo charcoal expressed by g/cm^3;

m refers to the mass of the bamboo

4.2.7 热 值

（1）定 义

竹炭的热值指单位质量的竹炭完全燃烧后冷却到原来的温度所放出的热量。根据燃烧产物中水分存在状态的不同，又分为高位热值（Higher heating value，HHV）和低位热值（Lower heating value 或 Net calorific value，LHV 或 NCV）。高位热值（HHV）是指单位质量的燃料完全燃烧产生的燃烧热和产物中水分冷凝为 0℃的液态水时所释放热量的总和；低位热值（LHV 或 NCV）是指单位质量的燃料完全燃烧后产物中水分冷却到 20℃时，所放出的热量。燃烧技术中的各种计算，均采用实际可利用的低位热值；研究文献中常用的为高位热值。

（2）测试方法

可参照 ISO 21626—2：2020《竹炭第二部分：燃料用竹炭》试验方法测定。可采用微机全自动量热仪、氧弹式量热仪来测定竹炭热值。

（3）影响因素

竹炭热值的大小主要与炭化终点温度、固定碳含量、挥发分和灰分含量等因素有关。一般来说，竹炭的热值随着炭化终点温度升高而增加；固定碳含量越高，竹炭热值越大；挥发分和灰分含量越高，则竹炭热值越小。竹炭的高位热值比低位热值高约 0.2MJ/kg，如经 800℃炭化后，毛竹炭化料的高位热值为 31.4002MJ/kg，低位热值为 31.1942MJ/kg；龟甲竹炭化料高位热值为 30.9570MJ/kg，低位热值为 30.7510MJ/kg。

charcoal expressed by grams;

V_{pore} refers to the pore volume inside bamboo charcoal expressed by cm^3;

V_{ture} refers to the true volume of bamboo charcoal expressed by cm^3.

True density (actual density) of bamboo charcoal is defined as the ratio between the unit volume and dry specimen mass of a sample, which excludes the volume of the voids and pores of bamboo charcoal. It can be calculated by formula (4-3):

$$\rho_t = \frac{m}{V_{true}} \tag{4-3}$$

Where,

ρ_t refers to the bulk density of bamboo charcoal expressed by g/cm^3;

m refers to the mass of the bamboo charcoal expressed by grams;

V_{ture} refers to the true volume of bamboo charcoal expressed by cm^3.

(2) Factors affecting the density

The density of bamboo charcoal depends on the bamboo carbonization parameters, equipment, and raw materials(bamboo species, parts, and growth ages), it varies quite extensively: from 0.2 g/cm^3 to 1.2 g/cm^3.

(3) Meaurement

Whether irregularly-shaped or powdered, true density of bamboo charcoal can be determined by alcohol (ethanol) displacement using a solid densimeter.

Bulk density (apparent density) of bamboo charcoal powders can be measured using a funnel and a receiver (the kit is available at the market), that is, load dry bamboo charcoal powders into a funnel, carefully fill the receiver with a given volume (usually 100 cm^3). The ratio of weight difference of filled and empty receiver to volume is expressed as bulk density (apparent

4.3 竹炭的特性

4.3.1 多孔性

（1）比表面积与测定

竹炭的比表面积指单位质量竹炭表面积的大小，即 1g 竹炭的外表面积与所有孔隙内面积的总和。竹炭比表面积一般小于 400m²/g。

（2）测定方法

参照 GB/T 19587—2017《气体吸附BET 法测定固态物质比表面积》，置于气体体系中的样品，其物质表面（颗粒外部和内部通孔的表面积）在低温下发生物理吸附。当吸附到达平衡时，测量平衡吸附压力和吸附的气体量，根据 BET 方程式求出单分子层吸附量，以此计算试样的表面积。

测定仪器：AS—703 型比表面积测定仪，全自动快速比表面积及孔隙度分析仪（ASAP 2020 Micropore System），全自动氮吸附比表面积测定仪（3H—200 型），电子显微镜测定法（Electronic Microscopic Examination），氮气吸附测定法（Nitrogen Surface Area）等。

（3）影响因素

竹炭的比表面积大小同工艺条件（炭化温度、炭化时间）、竹种、竹龄等因素有关，主要影响因素是炭化温度、炭化时间，竹种、竹龄对其影响没有明显规律。

作者研究团队研究了炭化温度从 300~1000℃的竹炭，其比表面积呈现出“两头低中间高”的趋势，800℃竹炭比表面积达到 385m²/g，300℃的竹炭最小比表面积为 23m²/g。炭化温度在 800℃条件下，

density) of bamboo charcoal powders.

(4) Requirements for applications

Highly densed bamboo charcoal means high strength in use, which tends to be utilized in prepacked charcoal products, drinking water purification, mat and cushion, etc, but on the contrary, soil amendment, and fertilizer applications can also benefit from the low density ones.

4.2.6 Acidity (pH value)

(1) Definition

Acidity of bamboo charcoal can be expressed by pH value, which is technically measured by indicator paper or meter.

(2) Factors affecting the acidity

The value is related to carbonization parameters and bamboo materials(the part in particular), usually falls into a range of 4.0 - 11.0.

(3) Measurement

Weigh approx. 5 g bamboo charcoal sample (un-dried, screened by 60 mesh or above), transfer to a 100mL conical flask, add 50mL deionized water to mix and gently heat to boil for 5 min, balance the mixture and filter, measure the cool-down filtrate by pH meter (acidimeter) or indicator paper.

(4) Requirements for applications

Acidic soil amendment requires bamboo charcoal of high pH value, and vice versa, and high acidic products(pH<7) is usually used for ammonia adsorption, low acidity(pH ≥ 7) for drinking water.

4.2.7 Calorific value

(1) Definition

Calorific value (heating value) is the amount of heat released during the combustion of a specified amount of charcoal sample. The calorific value is the total energy released as heat when fuel sample undergoes complete

随着炭化时间的延长，比表面积先增后降趋势，炭化时间 2.0~3.0h 时，竹炭比表面积最大。如姜树海研究发现了 750℃的毛竹炭比表面积 357m^2/g，日本阪田佑作等研究发现炭化温度 850℃竹炭比表面积为 370m^2/g，作者研究团队研究发现了方竹炭的比表面积为 432.42m^2/g。

4.3.2 吸湿性

（1）吸湿和解吸

当竹炭量一定时，环境温度越高，吸湿量越低，说明随着温度的上升，竹炭的吸湿能力逐渐降低；吸湿量随着环境相对湿度的增加而增大。在一定温湿度范围内，湿度对竹炭吸湿性的影响比温度显著。

（2）影响因素

竹炭吸附性能主要同其本身孔隙结构及周围环境温度湿度有关。竹炭吸附性能用吸附量（平衡吸附量）来表征，其平衡吸附量同炭化工艺（炭化温度、炭化速度、保温时间）、粒径大小、比表面积、环境的温度湿度等相关）。

据作者研究团队研究了不同炭化温度的竹炭自然放在浙江农林大学（国家木质资源综合利用工程技术研究中心）的实验室内，时间从 1—10 月，结果显示 500℃的竹炭平均吸收率 8.6%，600℃为 8.8%，700℃为 11.3%，800℃为 11.5%。

据卢克阳等研究表明，竹炭样品在环境温度为 15℃、相对湿度为 95%、炭化温度分别为 700℃和 1000℃的竹炭的平衡吸湿量最高，均达到 0.138g/g，即每克竹炭可以吸收 0.138g 水蒸气。在同等条件下，竹炭粒径对其吸湿量影响很小，竹炭的吸湿

combustion with oxygen under standard conditions. There are two expressions: higher heating value (HHV) and lower heating value (LHV, or equivalently net calorific value, NCV) which defined by the phase of water generated during combustion.

Higher heating value (HHV) is the amount of heat released during the combustion of a specified amount of it, which takes into account the energy in water evaporated from the fuel as it is combusted.

Lower heating value (LHV) is the amount of heat released by combusting a specified quantity and returning the temperature of the combustion products to 20 ℃ , which does not cover the latent heat of vaporization of water in the combustion products. By definition the higher heating value is equal to the lower heating value with the addition of the heat of vaporization of the water content in fuel (bamboo charcoal), the HHV of bamboo charcoal is averagely 0.2 MJ/kg higher than LHV of bamboo charcoal reckoned with experimental data.

(2) Measurement

The calorific value can be measured by microcomputer automatic calorimeter or oxygen bomb calorimeter, in accordance with local, regional, or international stardards, such as ISO 21626—2: 2020 *Bamboo charcoal—Part2: Fuel Applications*.

(3) Factors affecting the calorific value

Combustibility of bamboo charcoal can be explained by calorific value (heat releasing), which depends on its chemical components that is in close relation to carbonization temperature. As a rule, the higher temperature bamboo is carbonized, the higher is the fixed carbon content and calorific value of a bamboo charcoal. Generally the calorific value of bamboo charcoal ranges from 27.0 MJ/kg to 32.0 MJ/kg.

量与其比表面积呈正相关，但并非简单线性关系。

日本学者川健次等用日本高野竹炭进行了 250 h 吸湿性变化研究，发现在温度 40℃、相对湿度 90 % 条件下竹炭的平衡吸湿量可达 14.0 %（0.14 g/g）。

（3）湿度调节

在恒定相对湿度环境中，不同竹炭样品的吸湿率是不同的，也就是调湿能力是不一样的。一般情况下，竹炭具有吸附其重量的 1%~4% 的水的能力，吸湿率达 14%（0.14g/g，相对湿度 95% 时）。一般地，在房间里预先放入 7~26kg 的商品竹炭，可以调节 $30m^3$ 的空间到适宜的湿度（25℃，调节相对湿度 60%~90%）。当室内空气相对湿度过低时，竹炭又会解湿，增加空气湿度，可使室内保持人体最舒适的相对湿度。

4.3.3 吸附性

（1）原　理

竹炭吸附性是指竹炭表面吸收周围介质中其他物质的分子（如各种无机离子、有机极性分子、气体分子等）的性能。常用吸附速率和吸附量来表示。

竹炭具有大孔、中孔和微孔不同的孔隙结构、较大比表面积、分子不具极性，且不易与其他极性分子相结合，从而对大多数气态污染物，水、土壤中等有害物有较高的吸附能力，吸附方式包含物理吸附（物理作用）和化学吸附（化学反应）。

（2）提高法

竹炭吸附性能同其孔隙结构、比表面积等有关，所以可通过控制竹炭的生产工艺、修饰竹炭的孔隙结构等方法来提高竹

4.3 Functional performance

4.3.1 Porosity

(1) Specific surface area and determination

Definition: Specific surface area (SSA) is the total surface area, including the interstitial surface area of the voids and pores, of bamboo charcoal per unit of mass (expressed by m^2/g). Specific surface area of a typical bamboo charcoal is less than 400 m^2/g.

(2) Determination

Instrument: SSA is generally characterized by a BET Surface Analyzer, Micropore Physisorption Analyzer, based on Electronic Microscopic Examination, Nitrogen Surface Area (BET) method.

Determination: Due to practical reasons the adsorption of Nitrogen at a temperature of 77 K (liquid nitrogen) has been established as the method for the determination of specific surface areas. By means of the BET equation the amount of adsorbed gas, which build up one mono-layer on the surface, can be reliably and comparably calculated from the measured adsorption isotherm. The amount of molecules in this mono-layer multiplied by the required space of one molecule gives the BET surface area. More specifications can be found in ISO 9277 *Determination of the specific surface area of solids by gas adsorption using the BET method.*

(3) Factors affecting SSA

SSA of bamboo charcoal is associated with bamboo carbonization parameters (temperature and time), weakly relevant to bamboo species and growth ages.

The SSA of bamboo charcoal gradually reached to the maximum then declined, for the meso bamboo carbonized at 300-1 000℃, that is, highest around 800℃ with a SSA of

炭的吸附能力。常用的方法有活化法、改性法等。廖鹏等用氢氧化钾（KOH）对竹炭进行活化，制得以微孔为主的活性竹炭，BET 比表面积随磷酸与竹炭质量比的增大而增大，从 $300m^2/g$ 增大到 $2500m^2/g$。周建斌等利用纳米二氧化钛对竹炭进行改性，改性竹炭对甲苯的净化能力明显高于未改性竹炭。作者研究团队研究发现了微波改性竹炭（BC-MW）对水溶液中亚甲基蓝（MB）和酸性橙 7（AO7）的吸附比竹炭强。

（3）影响因素

竹炭吸附性主要与炭化工艺、吸附工艺及竹炭自身形状等因素有关。

①炭化工艺

炭化工艺直接影响竹炭的微观结构和比表面积，从而影响其吸附性能。在炭化温度低的情况下，竹炭以物理吸附为主，物理吸附是由范德华力结合的，作用力较弱。随着炭化温度的升高，物理吸附逐渐减弱，化学吸附逐渐增强，吸附量随温度的上升而增大。Saito 等研究发现不同炭化温度（600~1000℃）的竹炭对甲醛都有很强的去除能力，高温炭化的竹炭吸附甲醛后不易脱附，低温炭化的竹炭对甲醛的吸附能力弱。Asada T 发现竹炭对苯、甲苯、吲哚、粪臭素的吸附能力随炭化温度的升高而增强，对氨气的吸附能力随炭化温度的升高而下降，但对甲醛的吸附能力随炭化温度的变化没有明显的变化。

②吸附工艺

竹炭吸附时间是吸附工艺的一个最重要的因素，随着吸附时间的延长，竹炭的吸附性能也有所变化。肖继波等利用竹炭对染料的吸附性能研究也有相同结果，并

$385m^2/g$, and lowest at 300℃ with a SAA of $23m^2/g$. Carbonizing around 800℃ , the SSA of bamboo charcoal started to see a plateau at a time span of 2-3 hours, further carbonization did not provide a superior result of porosity. Diverse SSA results were reported, such as $357m^2/g$ (750℃ , *Phyllostachys edulis*), $432m^2/g$ (800℃ , *Chimonobambusa quadrangularis*), and $370m^2/g$ (850℃).

4.3.2 Hygroscopicity (moisture absorption)

(1) Absorption and desorption

Hygroscopicity (moisture absorption performance) is the phenomenon of attracting and holding water molecules via either absorption or adsorption from the surrounding environment(ambient air) at normal or room temperature. The Hygroscopicity of bamboo charcoal can be evaluated by weight or moisture content change of bamboo charcoal in a certain conditions, take a dry ambient air (with low humidity) for example, it will desorb (releasing water), and vice versa, which known as sorption hysteresis, on the ground of the principle, bamboo charcoal can be considered as a natural humidifier/dehumidifier working indoor.

In addition, ambient humidity outweighs temperature on hygroscopicity (equilibrium adsorption capacity) in a certain temperature range.

(2) Factors affecting hygroscopicity

Hygroscopicity of bamboo charcoal is relative to its performance limit, ambient temperature and humidity, it can be expressed by equilibrium adsorption capacity, which is also analogously relevant to carbonization parameters (temperature, time), particle size distribution, specific surface area and ambient temperature and humidity.

The Hygroscopicity of bamboo charcoal by carbonizing at 500℃ , 600℃ , 700℃ and

发现吸附时间达 2h 时，竹炭对活性染料的吸附基本饱和，吸附过程已经接近平衡。

③竹炭粒径规格

竹炭的粒径对竹炭的吸附性能也具有一定的影响。徐亦钢等研究发现竹炭在 2,4- 二氯苯酚的浓度为 1mg/L，pH=6.4，竹炭用量为 0.20g 条件下对其吸附，结果表明竹炭粒径尺寸越小去除率越高。杨磊等对竹炭对甲醛蒸汽的吸附性能研究也表明竹炭粒径越小吸附性能越好。

4.3.4 导电性

（1）电阻率

竹炭的导电性通常用电导率（K）来表示，它是电阻率（ρ）的倒数，即 $K=1/\rho$，$\rho=RS/L$，单位 Ω · cm。竹炭的电阻率（ρ）是用来表示竹炭电阻值特性的物理量。电阻率越小，表示导电性越好。

（2）测定法

竹炭按形状及尺寸不同，可以分为筒炭、片炭、碎炭、颗粒炭、粉末炭。

筒炭和片炭的形状比较规则，采用直流低电阻测试仪测试竹炭电阻，游标卡尺测量竹炭尺寸，通过电阻率公式计算竹炭电阻率：

$$\rho=\frac{R \cdot S}{L} \qquad （式 4–4）$$

式中：ρ——电导率，Ω · cm；

L——竹炭长度，cm；

R——竹炭电阻，Ω；

S——竹炭横截面面积，cm^2。

碎炭、颗粒炭、粉末炭形状差异较大，磨成粉后，按照 GB/T 24525—2009《炭素材料电阻率测定方法》中 4.3 方法、YS/T 587.6—

800℃ is 8.6 %, 8.8 %, 11.3 %, and 11.5 %, respectively, according to an investigation carried out from Jan. to Oct in authors' laboratory in National Research Center for Comprehensive Utilization and Engineering of Wood-based Resource (Zhejiang A & F university).

The equilibrium adsorption capacity can reach up to 0.138 g/g at a temperature of 15 C and relative humidity of 95 % with a bamboo charcoal prepared at 700℃ or 1 000℃ measured, nonetheless, the particle size (distribution) saw no remarkable impact on the capacity even determined with same conditions. It can be determined that the equilibrium adsorption capacity showed positive correlation with specific surface area, but not simple linear function.

A Japan group experimented on hygroscopic behavior of the bamboo charcoal (from Takano), it was found out that equilibrium adsorption capacity can achieve to 14 % (0.14 g/g) at a temperature of 40℃ and relative humidity of 90 %.

(3) Humidity conditioning (humidifier)

Moisture absorption performance usually differentiates on a certain ambient humidity. A typical capacity of water absorption can be determined at 1-4 % of the weight of bamboo charcoal applied, and its hygroscopicity can be reached at 14 % (0.14 g/g at RH 95 %). to put it another way, on condition that 7-26 kg of bamboo charcoal were placed in a room with 30 m^3, the humidity could be comfortably adjusted to 60 % (from 95% at 25℃), meanwhile it can discharge the once-absorbed moisture against dry weather.

4.3.3 Adsorption performance

(1) Principle

Bamboo charcoal's adsorption, indicated by adsorption rate or adsorption capacity,

2006《粉末电阻率的测定》测试竹炭。

具体测试方法：称取15~20g试样加入试样筒中。将带有活塞的试样筒置于试验机内，然后施加压力3Mpa。使用长度测量装置，测量样品高度 h。连好电线并接通电源。调整稳定电源电压，使通过试样的电流为500±0.02mA，然后测量试样电压降。测定两次，每次都用新的未测试过的试样。按以下公式计算电阻率：

$$\rho=\frac{S\cdot U}{Ih} \quad \text{（式 4-5）}$$

式中：ρ——电阻率，μΩ·cm；

I——电流，A；

h——试样颗粒柱的高度，mm；

S——试样筒容器的表面积，mm^2；

U——电压降，mV。

（3）导电原理

竹炭导电机理目前主要有从石墨微晶的形成和π电子理论、离子移动等角度进行探究。

①石墨微晶的形成和π电子理论

作者研究团队研究竹炭的导电性时发现，当炭化温度升高至800℃时，竹炭出现了碳碳双键与苯环共轭的吸收峰，导电性能增强，在温度达1100℃以上有类似石墨化结构趋势。赵丽华研究也发现随着炭化温度的升高、炭化时间的加长竹炭石墨含量增加，竹材高温炭化后碳原子价层中s与p轨道进行杂化，形成sp2杂化轨道，这时碳分子外层残留的2pz轨道，与相邻原子碳原子的2pz轨道便相互交叠形成大π键，形成类似于石墨的结构，因此增强了竹炭导电性（图4-2）。

means the capability of bamboo charcoal surface to absorb other molecules (inorganic ions, polar organic compounds, gases, etc.) in the surrounding medium.

Bamboo charcoal has different pore structures, large specific surface area, and non-polarity which does not tend to combine with other polar molecules. All these endow bamboo charcoal with the capability of absorbing most of the gaseous pollutants and contaminants from the water or soil. Adsorption could be physisorption and chemisorption.

(2) Improvement

Adsorbing performance of bamboo charcoal is influenced by pore structure and specific surface area. So it's feasible to improve the adsorbing capability of bamboo charcoal by controlling the production process or modifying the pore-structure.

Activation and modification can make bamboo charcoal great again. Micropores-dominated bamboo based activated carbon can be developed by potassium hydroxide(KOH) activating method, its specific surface area increased tremendously to 2 500 m^2/g, contrasted with the original area of 300 m^2/g, when more phosphoric acid was used in activation. Widely acknowledged that nano-scale TiO_2-doped bamboo charcoal shows a superior capability of benzene eliminating. Author's group also reported a microwave modified bamboo charcoal with improved adsorption of methylene blue and acid orange 7, which was far better than that of the untreated ones.

(3) Factors affecting adsorption

The adsorption of bamboo charcoal is mainly related to carbonization parameters, adsorption process, and physical form of bamboo charcoal.

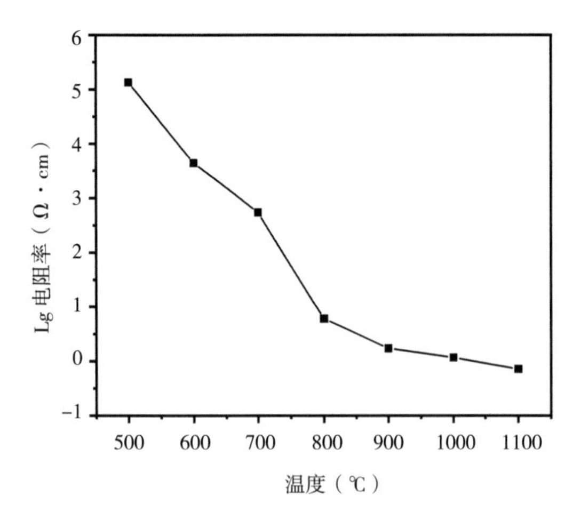

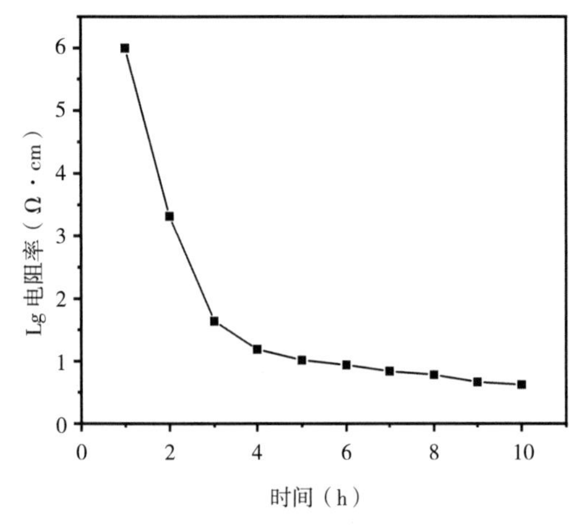

图 4-2　竹炭电阻率和炭化温度、炭化时间的变化关系

Fig4-2　The effect of bamboo carbonization temperature and time on resistivity of bamboo charcoal

②离子移动

作者研究团队研究发现炭化温度低于900℃时竹炭导电主要由离子移动引起的，竹炭灰分中的一些可溶性的钾、钠硫酸盐，不溶性的硅酸盐及少量磷酸盐等物质含量增加，促使离子的移动产生极化；同时灰分中一些金属元素掺和在竹炭中也会增强导电性。

（4）影响因素

竹炭导电性主要同竹炭微观结构、炭

① Carbonization

Carbonization parameter is of significance for producing bamboo charcoal with differed micro-structure and specific surface area, which inevitably determined its adsorption performance. Low temperature carbonized bamboo charcoal is prone to absorb through physisorption, bonding molecules with Van der Wals forces and thus a weak adsorption (to formaldehyde and other toxic gases), for bamboo charcoal made by increased carbonization temperature, chemical adsorption rise and physical adsorption fall, the adsorption capacity increases with the carbonization temperature rise. Bamboo charcoal carbonized at 600-1 000℃ demonstrated sound absorption performance can be obtained for all charcoals but desorption is difficult and arduous for higher temperature carbonized products.

According to a research by Asada et al, the adsorption capacity of bamboo charcoal for benzene, toluene, indole and scatol (3-methyl-indole) could be enhanced with the increase of bamboo carbonization temperature, yet for ammonia, it dropped when the carbonization temperature was high, it also presented a result that carbonization temperature appeared no notable effect on capacity for formaldehyde removal.

② Adsorption

Adsorption time span is the most important factor in adsorption. The adsorption performance would change with the absorption time. Similar conclusion was claimed by adsorption on reactive dyes using bamboo charcoal, that is, adsorption on dyes deemed to be stagnated in 2 h, which could be referred as an equilibrium.

化温度、化学组分等有关。

江泽慧研究发现竹炭的导电性能与微观结构相关，随着炭化温度的升高，竹炭维管束外鞘变得致密、光滑、平整，同时细胞间隙变小，因此密度增大，竹炭电阻性增强。

竹炭的碳含量与导电性能也相关，尤其是碳元素含量，Gabhi 在 950℃下热处理生物炭 8h，生物炭的含碳量从 86.8% 增加到 93.7%，生物炭电导率从 0.16s/m 增加到 106.41s/m，电导率提高了 600 多倍。

作者研究团队研究发现随着炭化温度的升高，竹炭电阻率呈下降的趋势，在 700~800℃间竹炭电阻率会发生显著下降。在同一炭化温度下，随炭化时间的延长竹炭电导率逐渐提高。炭化温度在 750℃附近是个临界点，前后电阻率相差约 50 万倍。

（5）性能提高法

①提高炭化温度法。通过提高竹材炭化温度可促进竹炭结构石墨微晶的形成来提高竹炭导电性。

②竹材催化炭化法。通过铁钴镍等过渡金属盐对竹材进行催化炭化，促进竹炭在较低的炭化温度条件下使竹炭结构向石墨化方向转变，形成石墨化竹炭从而提高竹炭导电性。

③金属元素迁移法，负载银（Ag）、铜（Cu）金属盐，使金属离子迁移至竹炭中，再经炭化形成“金属 / 金属氧化物包覆石墨化结构”的导电竹炭。作者研究团队研究了以竹炭为载体吸附硝酸铜 $Cu(NO_3)_2$ 后在热解作用下分解成氧化铜，继而同碳（C）反应生产单质铜（Cu）依附在竹炭上，电阻率达到 0.01Ω·cm，从而提高

③ Particle size

Adsorption performance of bamboo charcoal is certainly influenced by its particle size. Using 0.20 g of bamboo charcoal to adsorb 2,4-dichlorophenol at a concentration of 1 mg/L and pH at 6.4, the results turned that the small particle sized bamboo charcoal, with higher specific surface area, can promote pollutant elimination efficiency. And same result can be found when bamboo charcoal was applied as formaldehyde absorbent.

4.3.4 Electrical conductibility (resistance)

(1) Definition of electric resistivity

Electric resistivity (ρ) is used to refer to the conductivity of a substance, usually the lower the resistivity is, the better the conductivity is.

It can be classified as tubular bamboo charcoal, bamboo charcoal flakes, bamboo charcoal granules, and bamboo charcoal powder according to the shape and size of bamboo charcoal.

Resistivity of tubular, and flake bamboo charcoal can be calculated by formula (4-4):

$$\rho = \frac{R \cdot S}{L} \tag{4-4}$$

Where,

ρ refers to the resistivity, expressed by Ω·cm;

L refers to the length of bamboo charcoal, expressed by cm;

R refers to the resistance, expressed by Ω;

S refers to the cross sectional area of bamboo charcoal, expressed by cm^2.

(2) Measurement

The irregularly-shaped charcoals can be ground into powder, the resistivity of bamboo charcoal shall be measured in accordance to China's standards: GB/T 24525—2009 *Method*

其导电性。

（6）竹炭与其他材料的电阻值

据研究表明，不同材料类的电阻率数值如下：

竹炭电阻率 1×10^{8}~10^{0} Ω · cm；

针叶材炭（雪松、冷杉、马尾松）1×10^{8}~10^{-1} Ω · cm；

阔叶材炭（橡木、橡胶树、桦木）1×10^{7}~10^{0} Ω · cm；

硬阔叶树炭（白炭）1×10^{-2}~10^{-1} Ω · cm；

金属电阻率一般低于 1×10^{-5} Ω · cm；

碳材料电阻率一般低于 1×10^{-4} Ω · cm。

4.3.5 屏蔽性

（1）电磁屏蔽性能

一般来说电阻率越小，导电性越好，其屏蔽效果越佳。竹炭可以作为屏蔽材料主要原因在于：一方面竹炭像金属材料一样，具有较高的导电性，可反射电磁波；另一方面竹炭内部存在大量孔径不同的孔隙结构，其内反射作用有利于提高电磁波吸收性能，因此竹炭常作为可再生、低成本材料而应用于电磁屏蔽领域。国内外一些研究人员对竹炭的屏蔽性能进行了研究，结果见表 4–1。

（2）测试方法

竹炭及竹炭复合材料的屏蔽效能测试标准：参照 GB/T 30142—2013《平面型电磁屏蔽材料屏蔽效能测量方法》。仪器：矢量网络分析仪；方法：同轴法、波导法，将块状或者粉末状样品制样后置于矢量网络分析仪中检测。

for determination of specific resistance of carbon materials, or YS/T 587.6—2006 *Calcined coke for prebaked blocks—Testing methods—Part 6:Determination of electrical resistivity of granules.*

Transfer 15-20g of powdered bamboo charcoal samples to the sample cylinder of resistance tester, and apply pressure of 3 Mpa to press, measure the sample length in cylinder. Properly wire and power on, adjust the voltage to stabilize the current at 500±0.02 mA, measure the voltage drops. Calculate the resistivity by formula (4-5):

$$\rho=\frac{S\cdot U}{Ih} \tag{4-5}$$

Where,

ρ refers to the resistivity, expressed by μΩ•cm;

S refers to the surface area of the sample container, expressed by mm^2;

U refers to the voltage drop, expressed by mV;

I refers to the current, expressed by A;

h refers to the height of the sample in sample cylinder, expressed by mm.

(3) Conductive mechanism

The conductive mechanism of bamboo charcoal is mainly focused on the formation of graphite micro-crystals and π electron theory, and ion migration.

① Graphite crystallite formation and π electron theory

According to researches by author's group, conductivity of bamboo charcoal prepared from carbonizing at 800℃ , strengthened intensely, and conjugated absorption between carbon-carbon double bond and benzene ring, were observed from infrared spectrum, graphitized structure in bamboo charcoal was also obtained when bamboo carbonized beyond 1 100℃ . Reported that with the increase of carbonization

4.3.6 红外发射性能

（1）概　述

把能通过大气的红外线划分为三个波段，其中波长 4~1000μm 的红外线亦称为远红外线。但在实际应用中，常采用发射率来表征物质远红外线辐射能力。发射率可分为半球发射率和法向发射率（ε），目前国际上采用法向发射率来衡量产品的远红外性能。远红外线具有温热效应，可让身体保温及促进末梢血液循环，因此具有养生保健的功效。

在 8~14μm 红外波长范围内，测试温度 25 ℃时，竹炭法向比辐射率介于 0.800~0.960。

（2）测定方法

竹炭经过粉碎机粉碎并取过 100 目（150μm）标准筛的样品，烘至绝干，采用压片法制样。采用双波段发射率测定仪测定竹炭远红外发射率的测试温度为 25 ± 0.5℃，在 8~14μm 波段对样品进行多次测定后取平均值。相对标准偏差（RSD）不大于 1.00%，若不符合则需重新制样测试。

（3）影响因素

竹炭远红外发射率大小同理化性能、粒径大小、竹种、竹龄等因素有关。

理化性能：随着含水率和固定碳的升高，表现为先升高后平缓的趋势。

粒径：粒径较大时，使测试竹炭表面粗糙，导致竹炭测试的发射率偏大；粒径变小时，发射率随之降低，当粒径小于 150μm（100 目以上）时，竹炭远红外比发射率变化不显著且趋于稳定。

竹龄：作者研究团队选用 2~13 年的毛

temperature and time, graphite content in bamboo charcoal increased correspondingly. The s and p orbitals of carbon atom are being combined to form sp^2 hybrid orbital, meantime the π bond occurs between the remaining $2pz$ orbitals in the outer layer of the adjacent carbon atoms, thus a structure comparable to graphite is ultimately developed, which contributed mostly to the conductivity of bamboo charcoal (Fig 4-2).

② Ion migration

Ion migration is also the reason for the conductivity of bamboo charcoal prepared with carbonization temperature below 900℃ . Ion migration deduced polarization as well as trace metals in bamboo charcoal generates conductivity. Ions consist of various soluble potassium and sodium sulfates, insoluble silicates and phosphates.

(4) Factors affecting conductivity/resistance

The conductivity of bamboo charcoal is mainly relative to its micro-structure, carbonization temperature, chemical composition.

Electrical conductivity of bamboo charcoal was affected by its microstructure according to a study by Jiang group. With the increase of carbonization temperature, the outer sheath of the vascular bundle condensed and smoothed, and the cell gap narrowed, hence a higher density and better conductivity of bamboo charcoal were obtained. Reported by Gabhi, the carbon content is also a factor affecting the conductivity of bamboo charcoal. Carbon content of biochar, prepared at 950℃ for 8 h, increased from 86.8 % to 93.7%, and its electrical conductivity improved up to 106.41 S/m, a value 600 times higher than that of the untreated.

Author’s group found that the decline of resistivity of bamboo charcoal occurred as bamboo carbonizing temperature rise, a sudden

竹为原材料，经 800℃炭化烧制成毛竹炭，毛竹炭远红外发射率在 0.829~0.877，4~9 年达到最大，其值为 0.877。

竹种：散生竹竹炭的远红外发射率基本都大于丛生竹，主要原因是散生竹的固定碳含量大于丛生竹。

（4）提高方法

对竹炭进行加热、微波及高温再炭化处理，可以不同程度提高竹炭的远红外发射率，高温再炭化处理对于增强竹炭远红外发射率效果最为显著。作者研究团队通过研究表明，竹炭高温再炭化（热处理）可以提高其远红外发射率，处理以 800℃为宜，远红外发射率达到 0.95，竹炭远红外发射率主要受其固定碳含量影响，比表面积和孔径也会对其产生一定影响。

4.3.7 食用性

（1）可食性

根据 GB 28308—2012《食品安全国家标准 食品添加剂 植物炭黑》中规定，植物炭黑是以植物树干、壳为原料，经炭化，精制而成的黑色粉状微粒，无臭、无味，不溶于水和有机溶剂，是一种不溶性的着色剂，主要用于冷冻饮品、糖果、大米制品、小麦制品、糕点、饼干等。

依据 GB 28308，使用竹子为原料可以生产食品添加剂“植物炭黑”，企业应取得“植物炭黑”食品添加剂生产许可证，生产产品应符合相关食品安全标准要求。

从原材料来源的角度，竹炭粉属于植物炭黑，可作为食品添加剂，但从普通竹炭生产工艺过程来看，不能直接食用，需经过严格控制工序处理，去除碱溶性物质

drop in resistivity (strengthened conductivity) can be witnessed at a carbonizing temperature ranging from 700℃ to 800℃ , a inflection point which could be differentiated over 5 orders of magnitude before and after it, nonetheless, the resistance descends slowly and gently as the continued rise of carbonization temperature, as seen in Fig 4-2.

(5) Improvement

① The conductivity of bamboo charcoal can be enhanced plainly by increasing the carbonization temperature, which promotes the formation of graphite crystallite in bamboo charcoal;

② Catalytic carbonization is also a valid route to improve the conductivity of bamboo charcoal. Metals such as iron (Fe), cobalt (Co) and nickel (Ni) were used in catalytic carbonization of bamboo to drive the graphitization of bamboo charcoal even at lower carbonization temperatures;

③ Metal element migration is a process of loading metal salts into bamboo charcoal and subsequent carbonizing, produces a *metal/metal oxide@graphitized carbon/bamboo charcoal* composite, with conductivity optimized. Copper-modified bamboo charcoal was fabricate by thermal decomposition of copper nitrate [$Cu(NO_3)_2$] absorbed bamboo charcoal, the measurement indicated that its resistivity was as low as 0.01 Ω·cm, which implied an improved conductivity.

(6) Reference

Electrical resistance of common materials is collected for readers' reference:

Bamboo charcoal: 1×10^8 - 10^0 Ω·cm;

Softwood (cedar, fir and masson pine) charcoals: 1×10^8 - 10^{-1} Ω·cm;

Hardwood (oak, birch, rubber wood) charcoals: 1×10^7 - 10^0 Ω·cm;

White charcoal: 1×10^{-2} - 10^{-1} Ω·cm;

和残留焦油等。若作为食品添加剂，须严格控制砷、铅、镉、汞等重金属元素的含量，并必须严格按照国家标准生产，从原料、人员、设施设备、生产过程、包装运输、质量控制等均要达到卫生质量要求。

中国竹炭主要有筒炭、片炭、颗粒炭和粉炭等类型，部分类似植物炭黑，可作为食品添加剂，但不能直接饮用，可以食品接触用炭，比如饮用水净化、煮饭、煲汤等。

在医学界曾有过使用活性炭的案例。浙江医学院附属第二医院急诊科主任张茂说："活性炭在急救医学里是一种解毒剂，特别是一些口服的毒物，只是进入胃肠道，不会被血液快速吸收的有毒有害物质，我们会考虑使用活性炭片剂来解毒""除草剂

表 4-1　竹炭的屏蔽效能

Table 4-1　electromagnetic shielding performance of bamboo charcoals

材料 Material	频率 Frequency	电磁屏蔽 /dB EMI Shielding/dB
砖土窑竹炭 Bamboo charcoal (brick kiln)	10.0~3000.0MHz	21
日本竹炭 Bamboo charcoal (Japan)	10.0~3000.0MHz	23
≥ 750℃竹炭 Bamboo charcoal (carbonized ≥ 750℃)	4.0GHz	30
	35.0GHz	≥ 60
改性竹炭 Modified bamboo charcoal	0~3.0GHz	24
竹炭复合材料 Bamboo charcoal composite	10.0~1000.0MHz	45~75

Metals: ≤ 1×10^{-5} Ω·cm in general;

Carbon materials: ≤ 1×10^{-4} Ω·cm in general.

4.3.5 Shielding effectiveness

(1) Shielding

The electromagnetic interference (EMI) shielding effectiveness is better when a low electrical resistivity (high conductivity) material is applied. Bamboo charcoal of fine conductivity not only reflect electromagnetic wave, but also absorb it for its unique micro-pored structure, which can act as a renewable, cost-efficient alternative to classical electromagnetic shielding materials.

(2) Measurement

Electromagnetic shielding performance of bamboo charcoal (composite inclued) shall be measured and calcaluted in accordance with method specified in local, regional or international industrial standards, such like GB/T30142 (China), ASTM D4935 (USA), KS C 0304 (Korea), or standards issued by International Electrotechnical Commission(IEC). The sample can be bulk or planar prepared from powders, and measurement is usually conducted on a vector network analyzer (VNA) using either flange coaxial method or wave guide method.

4.3.6 Far-infrared radiation (emission)

(1) Far-infrared emission

Theologically infrared rays can be divided into three bands: near infrared (NIR, 0.76-2 μm), mediate infrared (MIR, 2-4 μm), and far infrared (FIR, 4-1000 μm). In the actual application, emissivity is often used to refer to the effectiveness in emitting energy as thermal radiation. Due to the fact that the rays' heat effect, it is ubiquitously used for warm-keeping purpose, which can facilitate blood circulation, achieving an aim of health care. The

中毒的人比较常见，急救医生会用这种片剂解毒，临床上已经验证了活性炭片剂可以解毒，而且活性炭片剂可以顺利排出体外，但是不能据此推论竹炭粉可以帮助排毒，因为活性炭片剂和竹炭粉是有区别的，需要临床试验。"

（2）安全性

可利用适量精制的竹炭粉末与淀粉、米粉、糖类等食品加工制成。在菲律宾、中国等地也开发竹炭花生（图 4–3）、竹炭面包，还有竹炭月饼、面条、巧克力、糖果等各种各样的食品。

图 4–3 竹炭花生和植物炭黑

Fig 4-3 Bamboo charcoal peanuts and bamboo-based edible carbon black

same phenomenon can be validated by testing on the pore structured bamboo charcoal.

Specific emissivity of bamboo charcoal is usually measured at a wavelength of 8.00 - 14.00 μm at 25 ℃ , the emissivity generally varies from 0.800 to 0.960.

(2) Measurement

Bamboo charcoal shall be crushed, ground or pulverized prior to screening by 100 mesh (150 μm), a disk sample can be obtained by pressure pressing of oven dried (25±0.5 C, 2 h) bamboo charcoal powders. Calculate the mean value after multiple measurements at a wavelength of 8-14 μm using a double-band emissivity tester. The relative standard deviation (RSD) shall be less than 1.00 %, otherwise, re-sample and remeasure it.

(3) Factors affecting far-infrared emission

The far-infrared emissivity of bamboo charcoal is related to particle size, and physicochemical properties, and the latter depends partly on bamboo species, and age etc.

Physicochemical properties: the FIR emissivity of bamboo charcoal tends to increase with the increase of moisture and fixed carbon content, respectively, it tends to level off when it reaches to the peak.

Particle size: the FIR emissivity could be presumably inaccurate or overestimated because of the high surface roughness originated from bigger particle that are not thoroughly ground and screened. Experimental findings demonstrated that bamboo charcoal particles no greater than 150 μm (100 mesh) can be optimal choice for far-infrared emissivity measurement.

Bamboo growth age: the FIR emissivity falls into a range of 0.829 - 0.877 for charcoals prepared from meso bamboo aged from 2 to 12 years, carbonizing at 800 ℃ . The emissivity

商家广告宣传的"竹炭面包等竹炭食品据说是可以净化血管""排毒养颜""竹炭可以吸附人体内有害物质，净化血液中的毒素，还有助于人体消化排泄，吸附多余油脂、调理肠胃、清洁肠道，具有排毒养颜的效果"等等。"食品有什么功能是很严肃的，需要大量的数据研究来支持，不能因为活性炭有吸附功能，就自然推断其变成食物后就有排毒功能了"。

"从来源的角度来说，竹炭粉也可以属于植物炭黑（图 32），可以称为食品添加剂。不过不是所有竹炭粉都可以称为植物炭黑，必须严格按照国家标准生产，从选材到加工工艺上，都符合标准才行。在使用上，竹炭的功用有很大的开发空间，日本等国家也在开发一些竹炭食品。国内在开发竹炭食品上，要慎重，是否能够直接加入，以及如何添加，都要好好研究"。

作为着色剂的添加剂，若声称具有特殊功用，如竹炭、植物炭黑具有排毒养颜功能，则须有大量的研究数据支持，这种研究可由企业进行，且研究必须获得官方专家组的认可，才能够称为功能。

据钟雨婷等的《食用竹炭粉的急性毒性与突变性研究》结果表明，微米级食用竹炭粉对大鼠的经口急性毒性属无毒级，对鼠伤寒沙门氏菌组氨酸缺陷型菌株和大、小鼠体细胞无致突变作用。

maximized (0.877) at 4-9 years, respectively. Paralleled conclusion can be drawn for fixed carbon content of bamboo charcoal that is closely related to bamboo growth age.

Bamboo species: the FIR emissivity of charcoal prepared from monopodial bamboo is generally higher than that of sympodial bamboo, the reason for that may be the fixed carbon content of the former is higher than that of the latter.

(4) Improvement

The FIR emissivity could be improved through reheating, re-carbonizing, or microwave processing. It turns out that high-temperature re-carbonizing is the most versatile route to the FIR emissivity enhancement. Re-carbonized at 800℃ , the emissivity of bamboo charcoal reached up to 0.95, a magnitude improvement contrasted with that of the untreated. The emissivity is mainly influenced by fixed carbon content of bamboo charcoal, and specific surface area and pore size distribution also make a difference in emissivity measurement.

4.3.7 Edible purpose

Vegetable carbon black is a black powdery particle made from plant trunk and shell, which is carbonized and refined, odorless, tasteless, and insoluble in water and organic solvents. The insoluble colorant mainly used in frozen drinks, candy, and bakery according to China's standard *GB 28308 (2012) National Food Safety Standard – Food Additive – Vegetable Carbon Black.*

In the view of the definition, bamboo can be a very raw material to produce food additive. As compulsory requirement by official departments, license on plant carbon black manufacturing shall be issued prior to production, all products, either bamboo charcoal as food colorant from manufacturer and distributor or end products from food supplier, shall meet relevant food quality and safety standards.

Easy to classify bamboo charcoal as a vegetable carbon black, yet, average bamboo charcoal is not readily to be edible, specific procedure and strict quality control shall be exerted on the processing, such as alkali soluble chemicals and residual tar in bamboo charcoal powers shall be removed, and heavy metals content limited. Production, packaging, and marketing of bamboo charcoal related food shall be in strict accordance with local and regional food quality and safety standards, laws and regulations.

At present, some enterprises are certificated to produce bamboo charcoal as food colorant, ensured full compliance with the good manufacturing practices in the Chinese mainland. Unique and featured bamboo charcoal food is commercially available in China and popular worldwide, e.g. bamboo charcoal peanuts, cookies, noodles, and other leisure foods.

Bamboo charcoal in China is mainly calssified as tubular bamboo charcoal, flake bamboo charcoal, granular charcoal and powdered bamboo charcoal, some of which are similar to vegetable carbon black, and some are for food contact purpose, such as applications in drinking water purification, and cooking, etc.

There have been cases of the use of activated carbon in medical profession, say, activated carbon is a useful antidote (detoxicating agent) in emergency medicine, provided that the swallowed toxics are not easily to spread into veins and only enter the gastrointestinal tract, activated carbon tablets could be an optional remedy to detoxicating. A newspaper reported that, people poisoned by herbicide are commonly found, and medics often use activated carbon tablet to detoxicate. It has been clinically proved that activated carbon tablet works in detoxicating, and it can be smoothly excreted from the body. However, it can not be inferred that bamboo charcoal powders can help detoxicating, because activated carbon tablets and bamboo charcoal powder are differential and further clinical trial is required. *(Morning Express)*

(2) Safety

In east and southeast Asia, refined bamboo charcoal has been used to develop, with flour, sugar and other ingredients to produce various foods (Fig 4-3).

Commercials that delivering information of blood purification, digestion benefiting, detoxification and beautification, is still controversial and shall be treated with caution. *Safety of edible bamboo charcoal (or bamboo base activated carbon) is a serious scientific issue, such detoxification function of bamboo charcoal foods can not be taken for granted, it could not be a foregone conclusion just base on its favorable absorption confirmed outside the body. (Morning Express)*

However, food additives as colorants claimed to be functional, like detoxification of bamboo charcoal or plant charcoal black, which should be provided with massive, supportive research data that can be carried out in industrial, moreover, results must be approved by the officials and professionals before disclosing to the general public.

Study stated that the acute toxicity (oral- rat) of edible bamboo charcoal powders can be graded as non-toxic, and no mutagenic effect was found on strains of histidine-deficient salmonella typhimurium, or somatic cells of rats and mice.

5　竹炭应用

Applications of Bamboo Charcoal

5.1 燃 料

竹炭本身的硫元素（S）含量比较低，热值介于 26.0~33.0MJ/kg，作为燃料前景非常广阔，是一种新型清洁能源，比传统煤燃料更清洁、卫生，环保和安全，是传统木炭替代品。可以竹炭粉为主要原料，添加淀粉胶或者羟甲基纤维素（CMC）胶黏剂混合，经成型干燥等工序制成的成型竹炭。成型竹炭是一种优良的烧烤、野炊燃料，在欧美也常作为壁炉用炭，在日本作为茶道专用炭，此外还可减少煤的用量，在工业领域广泛应用。

5.2 日常与健康

5.2.1 干燥与保鲜

（1）干 燥

竹炭具有吸湿性，可作干燥剂（图 5–1），一般平衡吸湿量为 14.0%；也可以同其他材料复合，形成竹炭基复合干燥剂，提高其吸湿性能，增强干燥效果。作者研究团队

图 5–1 干燥应用

Fig 5-1 bamboo charcoal prepacked desiccant for rice storage

5.1 Fuel application

Bamboo charcoal, which is a kind of fuel charcoal with a broad market prospect, is an excellent fuel for industrial and domestic use due to good calorific value (27.0-32.0 MJ/kg). Bamboo charcoal is cleaner and more environment friendly than coal because of its extremely low content of sulphur and nitrogen. Using bamboo charcoal as the substitute of traditional charcoal is a feasible way to protect timber resources.

Bamboo charcoal briquette, using starch or carboxymethylcellulose (CMC) as adhesive, can be served as quality fuel for cooking or barbecue (BBQ), often used in the fireplace in Western countries or the scenario of sado (tea ceremony) in Japan.

5.2 Daily health care

5.2.1 Desiccation and food preservation

(1) Desiccation

Bamboo charcoal can be served as desiccant (drying agent) as it shows a good moisture absorption (Fig 5-1), its equilibrium adsorption capacity generally reach to 14.0 %, can be considered as a natural humidifier/dehumidifier working indoor. It also can be chemically treated for enhanced adsorption performance. For instance, a calcium chloride ($CaCl_2$) modified bamboo charcoal composite stands out with improved adsorption capacity of bamboo charcoal and deliquescence (tendency to absorb moisture and dissolve in it) of calcium chloride, exactly, when the ambient temperature is 20±1℃, the adsorption capacity of the composite is determined to be

将氯化钙负载在竹炭中制得竹炭基复合干燥剂，克服单一竹炭吸湿量小和氯化钙易潮解问题。结果表明：竹炭基复合干燥剂吸湿性能比硅胶好，在相对湿度80%±1%，20±1℃的高湿条件下，吸湿量可达40%，在相对湿度40±1%，20±1℃的低湿条件下，吸湿量可达24%。

（2）保　鲜

竹炭具有吸湿和解吸特性，起到调节水分的作用。冯初国等研究得出竹炭能有效地控制茶叶水分的增加，且在茶多酚、咖啡碱的保存方面有较好的功效。结果表明，在8个月的贮藏期中，竹炭处理使茶叶的含水率增加率控制至31.47%，（空白对照54.61%），使茶叶叶绿素、茶多酚、咖啡碱、氨基酸的保留率得以不同程度提高，分别为空白对照的114.56%、106.70%、103.24%和111.81%。

竹炭还能吸附乙烯等特性，对蔬菜花卉、水果等的远距离运输保鲜有很好的效果，在冷藏箱、冰箱中作为保鲜剂。

5.2.2　洗护系列

竹炭具有吸附性（吸水吸油）、碱性、磨料（Abrasive）等功能特性，且竹炭的重金属含量基本满足食品安全国家相关标准（GB 5009.11、GB 5009.12、GB 5009.15 和 GB 5009.17）要求，所以竹炭材料能作为洗护系列的添加剂原料，目前已开发出竹炭香皂、洗发液、沐浴露、面膜、洗面奶、牙膏等洗护产品，可用于清洁污垢和油脂，这可谓“黑炭洗白脸”（图5–2）。

40 %, 24 % at a relative humidity (RH) of 80 %, and 40 %, respectively.

(2) Food preservation

Because of its hygroscopicity (absorption and desorption) mentioned in Clause 4.3.2, bamboo charcoal can be apply to moisture regulation. According to a research by Feng et al, it can effectively contain moisture increment in tea, and preserved tea polyphenols (Tp) and caffeine in tea, keeping its original flavor. In a 8-month tea storage, moisture content of bamboo charcoal accompanied sample was controlled at 31.47 % (54.61 % for a blank control), retention rates of chlorophyll, tea polyphenols, caffeine, and amino acids are also secured.

Ethylene, a ripening agent, can be absorbed by bamboo charcoal which makes bamboo charcoal a candidate preservative that is readily to serve in fruit and vegetable preservation in long distance transport, cold storage, or household refrigerator.

5.2.2 Personal care

Based on alkalic, abrasive, oleophilic and hydrophilic properties of bamboo charcoal, as well as its extreme low content of hazards of heavy metal contamination, bamboo charcoal powders can be applied as alternative to plastic or mineral additives (e.g. plastic microbead, TiO_2) in cosmetic and personal care products (Fig 5-2).

5.2.3 Far infrared health care

(1) bamboo charcoal

Far Infrared Rays are invisible waves of energy that have the ability to penetrate all layers of the human physical body, penetrating into the inner-most regions of the tissues, muscles and bone. It could gently heal, soothe,

图 5–2　竹炭洗护系列产品

Fig 5-2　bamboo charcoal as functional additives in daily chemicals

5.2.3　远红外保健功能

（1）竹炭原料

竹炭具有较高红外发射率，这就为竹炭在保暖、保健等领域应用提供理论依据。目前有许多相关产品问世，如：竹炭远红外护具、护腰带、护腕、汽车、凳椅等坐垫和保健包。

（2）竹炭纤维纺织品

据日本研究报道，日本生产出的竹炭纤维，与化纤、棉线等交织在一起，热量可增加 10%，并具有吸附性、透气性好等优点。竹炭纤维由于其良好的吸附性、抑菌性，可以制成抗菌毛巾、医疗防护服饰、口罩、婴幼及孕妇防护服、袜子；蓄热保暖，可以做成保暖内衣，防寒服；可以加工成各种规格的被褥、枕头、床垫等。还可在空调设备以及汽车座舱空气的过滤器中，应用竹炭纤维过滤材料，可吸附臭气及细菌并具有防毒作用。根据闫鸿敏等研究结果表明，其主要有以下性能：

①远红外发射性能。竹炭涤纶针织物具有远红外功能，在温度 25 ℃，波长

stimulate and detox the physical body, as well as the mind. A good far infrared emission was founded in bamboo charcoal, and it now was widely used in warm-keeping and therapeutic applications, such like pad, belt, bracer, cushion and health-care pack.

(2) bamboo charcoal fabric

Bamboo charcoal fiber can be inter-grated into synthetic fiber or cotton yarn, which was applied as textile fabrics outworks other fabrics in warm-keeping (10 % higher in heat accumulation), absorption, bacteriostasis, and air permeability, it also is used as raw materials for fabricating towels, personal protective equipment (e.g. masks, protective clothing), underwear garments, home textiles (e.g. pillow, cushion, mattress), and automobile accessories.

The features of bamboo charcoal fabric include, but not limited to:

① Warm-keeping: the far-infrared emissivity of bamboo charcoal is measured approx. 87% at 25℃ in a wavelength of 8-14 μm, a result indicating prospective applications in textile industry.

② Absorption: bamboo charcoal/terylene (polyester) composite fabric has a tendency to absorb smelly sulfur compounds (hydrogen sulfide, methanethiol, ethanethiol), and nitrogenous compounds (ammonia, amines), released from human metabolism. For example, ammonia absorption of the composite fabric can be up to 25 % at a time span of 20 min.

③ Permeability: hydrophilized bamboo charcoal/terylene composite fabric show a fine moisture transport and air permeability.

④ Bacteriostasis: bamboo charcoal/terylene composite fabric has a good rate of bacterium restraining (65%), compared to

8~14μm，法向发射率为 86.9%；

②吸附性能。竹炭涤纶针织物能吸附硫基化合物（硫化氢、甲硫醇、乙硫醇等）和含氮化合物。竹炭涤纶针织物对氨气的吸附在 20min 左右就已经基本达到吸附饱和，饱和吸附率约为 25 %。

③导湿性能。经亲水处理过的竹炭涤纶针织物的导湿性能比未经处理的要好。

④抑菌性能。竹炭涤纶针织物抑菌率为 65 %，说明竹炭涤纶针织物有一定的抑菌性能。

5.2.4 电磁屏蔽效能

（1）竹炭产品

导电性好的竹炭具有防电磁波作用，可放置电视机、电脑、手机类电器旁吸附阻隔电磁波，已开发竹炭屏蔽服、围裙、鼠标垫等产品。

（2）竹炭复合产品

以导电竹炭为原料同其他材料复合研发出导电竹炭新材料及轻质电磁屏蔽材料，用在电子仪器、船车、涂料及新型高级建筑材料上，减少电磁波对人体危害。如：K.H.Wu 等人通过竹炭与聚苯胺复合得到了电磁屏蔽材料，在频率为 2~40MHz，对 7.2GHz 和 33GHz 的电磁波的屏蔽效能分别为 8dB 和 17dB。

H. L. Jia 等人研究的竹炭聚酯纤维织物对 500MHz 的电磁波的屏蔽效能为 45dB。作者研究团队研究竹炭与乙烯—乙酸乙烯共聚物（EVA）复合制备了竹炭 /EVA 电磁屏蔽复合材料，在 22557.5 MHz 时 1~4mm 厚的板材屏蔽效能为 35.2~48.5dB。

barely presented antibacterial behavior of counterparts, which makes it an ideal choice for textile industry.

5.2.4 Electromagnetic shielding

(1) Bamboo charcoal

Based on electromagnetic shielding performance of conductive bamboo charcoal, it can be applied to manufacture bamboo charcoal garment, apron, and mat products for indoor use.

(2) Bamboo charcoal composite

Conductive bamboo charcoal based composite is of great advantages utilized in electronic devices, vehicles, painting and coating, and novel building materials, to reduce or avoid presumable electromagnetic hazards.

A bamboo charcoal/polyaniline composite demonstrated that the shielding effectiveness reaches to 8 dB and 17 dB for 7.2 GHz and 33 GHz, respectively, in a electromagnetic frequency range of 2 - 40 MHz. According to a study on bamboo charcoal/polyester fabric indicated that the shielding effectiveness is around 45dB for 500 MHz electromagnetic radiation, and authors' investigation showed bamboo charcoal/ ethylene-vinyl acetate copolymer (EVA) composite has a shielding effectiveness of 35.2-48.5dB for panels of 1 mm - 4 mm in thickness in a frequency of 22557.5 MHz.

5.3 Environmental protection

5.3.1 Water purification

(1) Drinking water

① Mineralized water

Residual chlorine and heavy metals in water could be absorbed by porous bamboo

5.3 环保领域

5.3.1 水质净化

（1）饮用水

①饮用水成矿物质水

竹炭能吸附饮用水中余氯和重金属离子，竹炭浸出液的pH较高可中和饮用水成为弱碱性水，还可溶解竹炭灰分中一些人体所需的微量元素如钾（K）、钙（Ca）、钠（Na）、镁（Mg）、铜（Cu）、锌（Zn）、铁（Fe）等，使饮用水变成矿物质水，味美可口。

用于净化自来水的竹炭因吸附饱和，建议1~2天内更换。

②循环利用次数

竹炭应用于煮水，持续煮水或循环利用，其饮用水微量元素指标也均符合生活饮用水卫生标准GB 5749—2006。随着竹炭煮水的循环使用，竹炭对铅（Pb）、铬（Cr）元素的吸附率增加，在煮水4次后，竹炭对钾（K）、钠（Na）、镁（Mg）、铁（Fe）等微量元素含量溶解量也有所降低，竹炭煮水至少可以循环使用4次。

③炭水使用比例

竹炭使用量一般为水重量的2%~5%。竹炭煮水可提高0~2℃自来水沸腾温度，对铅（Pb）、铬（Cr）元素存在明显的吸附效果，并会溶解出钾（K）、钠（Na）、镁（Mg）元素，其他元素在煮水过程中规律不明显，随着炭水比例的增加，竹炭煮水后自来水常量元素总含量明显增加。竹炭应用于自来水常温浸渍，随着浸渍时间的增加，微量元素总含量明显增加，前3h较明显，竹炭常温浸渍饮用水3h前饮用较佳，

charcoal, and water turns slight alkaline, water soluble elements, potassium (K), calcium (Ca), sodium (Na), magnesium (Mg), copper (Cu), zinc (Zn) and iron (Fe) can mineralize water, which gives health benefits. Bamboo charcoal for drinking water purification shall be replaced at intervals of one day or two.

② Service cycle

Even though beneficial elements can be found in bamboo charcoal purified drinking water, it ought to meet local or regional sanitary requirements for drinking water quality. Absorption to lead (Pb), chromium (Cr), and zinc (Zn) is gently increased as use counted. A 4-cycle use in boiling water, soluble minerals gradually descend, and the equilibrium adsorption of bamboo charcoal seems to be accomplished, which can be deemed as its full cycle of bamboo charcoal usage.

③ Usage

A temperature of 0-2℃ can be increased in water boiling by adding bamboo charcoal. Obvious adsorption of lead (Pb), chromium (Cr), and zinc (Zn), release of major elements in water can be attained with initial 3 h soaking that is mostly suitable for drinking, and addition of 2%-5% of bamboo charcoal, as well as under 30 minutes water boiling, is also advised.

(2) Cooking

Flake bamboo charcoal is more often used in cooking in east Asian than any other regions. High temperature (≥ 800℃) carbonized bamboo charcoal shall be washed and boiled for hygienic sake prior to soup or rice cooking (Fig 5-3). The features of bamboo charcoal applied in cooking:

The acidity (pH value) of bamboo charcoal soaked water can reach up to 10, and turns

图 5–3　专用竹片炭

Fig 5-3　bamboo charcoal for food contact (cooking) applications

沸腾时间应尽量不超过 30min。

（2）蒸煮用水

日本学者曾利用静冈市新间地区的 5 年生毛竹制备炭化温度为 800℃和 1000℃的竹片炭（图 5–3）。在沸水中加热 15min 杀菌消毒后取出，放入煮饭锅同米一起煮饭，竹炭的使用量是水的 10%。结果表明：

①竹炭浸出液的 pH 值较高，达到 10；加入米之后浸渍；再加热后，pH 值降至 8 左右，但是仍比只用水煮的饭 pH 值要高。用竹炭长时间浸渍在水中得到的竹炭浸出液煮米饭，更软。其中 800℃的竹炭上表现出更明显。

②用了竹炭的煮饭用液中，全糖和蛋白质的溶出较多。推测用了竹炭的煮饭用水，其高 pH 值促进淀粉的分解、抑制老化上产生了有效的作用，同时也能防止饭粒氧化而发黄，不产生馊味，米饭更香甜、更健康。

③竹炭的远红外发射作用使锅中的温度均匀一致，温度比平常提高 2~3℃，米芯充分熟透，提高了热效率。

（3）污水处理

①氨氮

王祝来等研究了竹炭对水溶液中氨氮的吸附特性，测定了不同竹炭粒径、溶液

mildly alkaline (around 8.0) when rice cooking is finished, which is a bit higher than regular water, nonetheless, the rice could be far more tasty than that of cooking with the tap water;

Protein and whole sugar leaching could be more when bamboo charcoal is used, it may be ascribed to starch hydrolysis and oxidation resistance of grains in alkali condition, which makes a healthy diet.

The heat may be much more evenly distributed and the temperature inside boiler is 2-3℃ higher owing to the infrared radiation emission of bamboo charcoal used during cooking, the food (rice) is thus cooked thoroughly for the improved thermal efficiency.

(3) Waster water management

① Nitrogen removal

Impact factors (particle size of bamboo charcoal, initial concentration, pH value of ammonia solution, and adsorption time) on Ammonia absorption performance had been explored and proved that: the adsorption is in accordance with Freundlich adsorption isotherm. The smaller particle size of bamboo charcoal is, the more obvious the adsorption performance shows. when pH value increases in acidic environment, the absorption would be apparently improved, and same pattern could be found as well when the initial concentration of ammonia solution or absorption time was considered. The equilibrium adsorption capacity could reach to 0.21 mg/g for a 6-h adsorption. Generally bamboo charcoal can be regenerated by alkali solution but with undermined performance, for instance, 64 % of original capacity can be attained at fourth regeneration.

Reported that a microbial community was successfully developed both on the surface and

初始氨氮浓度、pH 值、吸附时间对吸附效果的影响。结果表明：在酸性条件下，pH 值增大吸附量增加较快，pH 值为 7 时吸附效果最佳；在不同氨氮初始浓度下，竹炭吸附量随着浓度增大而急剧增加；吸附时间越长，吸附量越大，6h 时达到吸附平衡。竹炭饱和吸附量最高达到 0.21mg/g；竹炭颗粒粒径越小，吸附效果越显著。竹炭对氨氮的等温吸附符合 Freundlich 吸附等温方程式；用氢氧化钠（NaOH）溶液进行再生，再生次数越多吸附量显著下降，4 次再生后达到原吸附量的 64%。周珊等将经过高效复合微生物菌群固定在竹炭颗粒的表面和孔隙内部，制备固定化微生物竹炭，对废水中的主要污染物氨氮进行降解。

②重金属离子

张启伟研究了竹炭对饮用水中氟离子（F^-）、铅离子（Pb^{2+}）的有较强的吸附能力，比吸附量均随吸附时间的延长而增加，但在 90min 后均趋于缓慢。

作者研究团队发现竹炭对重金属离子吸附存在物理和化学吸附，其吸附性能与其比表面积、孔径及 pH 值等有关；炭化温度为 800℃的竹炭对汞离子（Hg^{2+}）、铅离子（Pb^{2+}）、铬离子（Cr^{6+}）吸附效果较好，炭化温度为 1000℃的竹炭对镉离子（Cr^{6+}）吸附效果较好。

刘创研究了竹炭对溶液中镉（Cd）(Ⅱ) 的饱和吸附量达到 11.0mg/g，初始浓度为 15mg/L 时，增加竹炭的用量，镉离子（Cd^{2+}）的吸附率增加，镉离子吸附率可达 99%。

此外，竹炭对废水中的汞、砷（Ⅲ）离子也都有较好的吸附效果。

inside the pores of the bamboo charcoal, which can be used to degrade ammonia-rich waste-water.

② Hazardous metals

A strong adsorption to F^- and Pb^{2+} in drinking water was confirmed by Zhang, the capacity of ions tended to rise sharply in 90 min and slower later.

Author's group also found that adsorption of metals (Hg^{2+}, Pb^{2+}, Cr^{6+}, Cd^{2+}) was highly affected by carbonization parameters, to put it another word, carbonization temperature gives diverse specific surface area, pore size and acidity, respectively, of the bamboo charcoals, results showed 800℃ carbonization is a good option to achieve optimal adsorption.

It has been reported that the saturated adsorption capacity to Cd^{2+} can be up to 11.0 mg/g, and the adsorption rate could secure a 99 % efficiency by an increased dosage of bamboo charcoal. Besides, a similar conclusion can be approached for mercury and arsenic removal using bamboo charcoals.

③ Organic compounds

The adsorption capacity to organic compounds in waste-water depends on particle size and dosage of bamboo charcoal, contaminant concentration, and adsorption equilibrium time. Experimental findings supported that bamboo charcoal provided an expected adsorption to various organic compounds (say 2,4-dichlorophenol, methylene blue, and acid orange 7), including pigments and dyes, in waste-water from paper-making and pigment industrial.

5.3.2 Air purification

(1) Bamboo charcoal

As a selective and sensitive absorbent, bamboo charcoal shows a large specific

③其他有机物

竹炭对污水中 2, 4– 二氯苯酚、染料废水、造纸厂排放流体中的色素和水溶液中亚甲基蓝和酸性橙等的吸附，试验表明竹炭具有较强的吸附能力，吸附效果主要与竹炭的粒径、用量和有机物的浓度以及吸附平衡时间等有关。

5.3.2 室内空气净化

（1）原竹炭类

竹炭作为常用吸附剂的一种，其具有选择性强、比表面积大、孔隙结构特殊且性能稳定等特点被广泛应用于室内空气净化。较常见的是竹炭被用于新装修的室内空气净化，用于吸附家具、地板以及合成建筑材料等所释放的甲醛、苯类化合物、挥发性总有机物（TVOC）以及乙酸丁酯等挥发性有机物，以达到室内空气净化的（图 5–4），比如市面上较为常见的竹炭包。如王晓旭等对市面上的四种竹炭研究发现：在相同温度下的吸附能力都是随着

surface area and porous structure which is of vital advantages to air purification indoor, especially the newly decorated. Affordable bamboo charcoal prepackages are frequently used to absorb formaldehyde and volatile organic compounds (VOCs, like benzene, toluene, and various esters) released from furniture, flooring, and other synthetic building materials (Fig 5-4). According to an assessment, better adsorption can be obtained from smaller sized bamboo charcoal particle/powder, and 60 mesh (0.250 mm) screened bamboo charcoal outperforms in absorption of toxic gases, however, it is prone to decline for the powder smaller than 80 mesh (0.180 mm), in addition to particle size, the capacity is still affected by temperature and time, it generally takes 4-5 days to replace bamboo charcoal used at 20-25℃ indoor that very equilibrium adsorption is roughly implemented.

Also, bamboo charcoal could be used for absorptive treatment to flue gas of coal combustion, from which mercury (Hg) and sulfur dioxide (SO_2) can be selectively absorbed before emission.

图 5–4 竹炭空气吸附应用

Fig 5-4 bamboo charcoal-based products for air quality improvement

试样粒径的减小先升高，过60目筛子的竹炭试样吸附率最高，但是随着竹炭粒径的继续降低到过80目筛子时，竹炭试样的吸附率反而降低了。除了竹炭粒径以外，竹炭的吸附性能还受到温度、吸附时间以及气体种类的影响。李倩研究发现竹炭在温度20~25℃之间对甲醛、苯、TVOC的吸附性能随竹炭的炭化温度升高呈现上升趋势，竹炭吸附去除有害物质的最佳持续时间在4~5d。除了室内空气净化外，竹炭在燃煤烟气净化中也有一定应用。如谭增强等利用小型燃煤烟气汞脱除实验台模拟烟气气氛，研究表明：竹炭对燃煤烟气中汞（Hg）和二氧化硫（SO_2）有一定的吸附能力，起到吸附脱除作用。

（2）负载改性竹炭类

在实际应用过程中，竹炭存在有吸附饱和以及反释放的情况，如果竹炭放置时间过长，吸附量达到饱和会重新释放有害气体，因此需要对竹炭定期处理。不少专家学者对竹炭进行改性处理的研究。有学者发现二氧化锰（MnO_2）对甲醛具有最高的反应性能，且主要产物为二氧化碳（CO_2），通过对竹炭负载锰氧化物来消耗竹炭吸附的甲醛来提高竹炭的吸附量；对竹炭负载光触媒体二氧化钛进行改性，在室温、光照条件下二氧化钛（TiO_2）与污染物接触可到达净化的作用，二氧化钛解决了竹炭吸附过程中的饱和及反释放的问题。此外通过对竹炭分别进行负载溴和氯化锌（$ZnCl_2$）化学浸渍的方法改性来提高竹炭对汞的吸附能力。

（3）竹炭复合类

张启伟等将二氧化硅（SiO_2）、竹炭和

(2) Modified bamboo charcoal

The occurrence of desorption turned out to be the rule rather than the exception in harmful gas management, it is necessary to regularly replace overtime used bamboo charcoal. Modification of bamboo charcoal can greatly improve its adsorption property, of which catalyst-doping is a common solution. MnO_2-modifed bamboo charcoal has a high reactivity to formaldehyde, it can continuously degrade the volatile to release harmLess CO_2 and water, spontaneously making room for absorbing, in time a cycle happens. Same benefit of toxic removal (heavy metal, noxious gas) can be achieved by TiO_2 or other chemical modified bamboo charcoal.

(3) Bamboo charcoal composite

Performance coating produced by bamboo charcoal, silica, titanium dioxide, and polymers, was applied to leather substrate, giving a solid adsorption and degradation of indoor or in-car contaminants or harmful gases. Inorganic Composite was prepared by sintering of bamboo charcoal and ceramics (diatomaceous earth, attapulgite and kaolin), and its capacity of adsorption to formaldehyde, ammonia and hydrogen sulfide was depicted as excellence: 87.7%, 94.6%, and 96.3%, respectively, provided that a catalyst was introduced or doped to the composite, the capacity will be reinforced, for instance, formaldehyde adsorption capacity of TiO_2-modified bamboo charcoal composite can reach up to 93.75%.

5.3.3 Soil amendment

The features of bamboo charcoal applied in soil amendment includes:high absorption of CO_2, imputrescibity (difficulty to be decayed), water retaining, fertilizer conservation, and alleviated soil acidification

纳米二氧化钛（TiO_2）制成功能性材料涂布在革基布上得到环保功能性产品，对居室或车内的污染物或有害气体有吸附和降解作用（尤其是光照条件下，吸附和降解效果更优）。韦冬芳等将竹炭、硅藻土、凹凸棒和高岭土按一定比例混合造粒后在氮气（N_2）气氛中高温烧制成有高强度、高吸附性能和高比表面积的竹炭陶复合材料，其对甲醛（CH_2O）、氨气（NH_3）和硫化氢（H_2S）的吸附率分别达到 87.7%，94.6% 和 96.3%。彭虎等以竹炭和紫砂土为原料混合球磨并无氧烧结制得强度高、化学稳定性好、适应性强的实用型竹炭笼芯紫砂毫或微球空气净化器，净化器对氨气（NH_3）、硫化氢（H_2S）、甲醛（CH_2O）的净化率分别为 92.6%、94.3%、87.7%。王喜华等人以高岭土、纳米二氧化钛改性竹炭为颜料制备的涂布纸能经受涂料配制中的 92.6%、94.3%，对甲醛的去除率可达 87.7%，且纳米二氧化钛改性竹炭涂布纸对甲醛的吸附降解率高达 93.75 %。

5.3.3 土壤改良

杉浦银治认为竹炭对二氧化碳（CO_2）烦的吸附性能高，不腐烂、保水、透水、保肥性好，可作为土壤改良剂使用（图 5-5）。姚远等研究认为，竹炭用于土壤改良能显著提高了蜘蛛类个体数量，降低了植食类昆虫的个体数量，马建伟等研究认为竹炭对土壤中镉的去除，会减少土壤中氟化物的可用性和其后的茶植物吸收氮。竹炭施加于土壤，起到减少化肥农药流失、保肥增肥、提高土壤的通气性、保水性、缓解土壤酸化之功效。

(Fig 5-5). According to authors' research, the physico-chemical properties of soil improved apparently by spreading granulated bamboo charcoal, with highly effective concentration of hydrolysable nitrogen (N), available phosphorus (P) and potassium (K), and exchangeable calcium (Ca) and magnesium (Mg), whereas lowered level in available zinc (Zn) and copper (Cu), in which *festuca arundinacea* cultivated was founded that both the growth and the germination of root and leaf were fully stimulated. Farming workforce was also benefited from the bamboo charcoal-amended soil, for example, bamboo charcoal applied in soil amendment, not only drived off cadmium (Cd) in soil, but also fluorides minimized drastically, nitrogen availability guaranteed for growing tea trees. And an interesting field investigation concluded that there was a rise of spiders in the amended field that directly projected by a fall of phytophaga (pests).

5.4 Construction decoration

5.4.1 Indoor humidity conditioning

(1) Flooring

Moisture absorption and desorption of

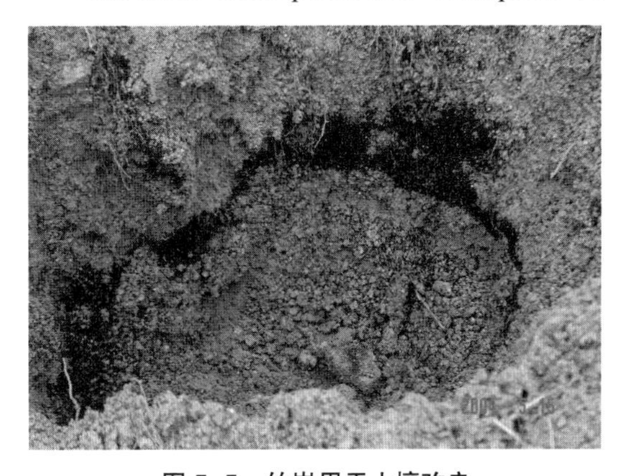

图 5-5 竹炭用于土壤改良

Fig 5-5 bamboo charcoal for soil amendment

作者研究团队研究了竹炭可改变土壤理化性质和促进高羊茅的生长，施过竹炭颗粒的土壤，其理化性能得到了改良，水解氮、有效磷、速效钾、交换性钙和镁等元素含量均明显提高，而有效锌和铜等金属含量则相对降低；竹炭对高羊茅的发根、发叶及生长也有不同程度的促进作用。

5.4 建筑装修

5.4.1 室内调湿

（1）地　板

竹炭有吸湿和解湿功能，起到调节环境湿度的作用（图 5-6）。当环境相对干燥（湿度较低），竹炭又会发生解吸，在日常生活中将竹炭置于地板底（地板调湿炭），可调节湿度，抑制霉菌、虫、微生物的繁衍，防止地板变形。

（2）装饰（墙）板

沈跃华等发明公开一种竹炭纤维调湿板，调节环境湿度稳定，使室内装饰产品不受潮、霉变或干燥开裂、变形等优点。李文彦等采用竹炭、硅藻土和黏土为原料制备出一种用于内墙装饰的含炭新型的室内功能装饰材料。吉行那穗子等利用竹炭、碎石粉和骨灰等材料制备竹炭基复合材料具有良好的调节湿度的功能。

5.4.2 室内除味

竹炭置于室内、橱柜中可以吸附装修材料释放的有害物质。杨磊等以通过竹炭为基料、热塑性白乳胶为黏接体、无纺布为增强和表面装饰材料进行研究，通过热

bamboo charcoal can make itself a natural humidity conditioner used indoor (Fig 5-6). When applied beneath the flooring, bamboo charcoal can absorb excess moisture to avoid microorganism and fungal growth, and possible deflection of flooring.

(2) Decorative (wall) panel

Humidity-conditioning Bamboo charcoal/polymer panel is invented for house (mainly walls) decoration. It asserted that it can control indoor humidity, diminish wetting, protecting decoration and furniture in damp season. Decorative panels were also made from bamboo charcoal and minerals, by report it showed versatile functions in humidity conditioning, toxic gas absorption, etc.

5.4.2 Deodorization

Place bamboo charcoal in room or closet is prone to absorb toxic gases and remove unpleasant smell (harmful gases actually) when furniture installed or house decorated

图 5-6　竹炭调湿功能的应用

Fig 5-6　bamboo charcoal for humidity conditioning

压胶合制备竹炭复合板材，具有吸附功能复合板材。结果表明：在 55℃、3h 的吸附条件下，竹炭对甲醛蒸气的吸附量最大，达 68.5mg/g。

in recent. Panel made from hot pressing of bamboo charcoal powder and resin, non-woven fabric enhanced and faced. The results showed that under the adsorption condition of 55 ℃ and 3h the adsorption amount of formaldehyde on bamboo charcoal was maximal up to 68.5mg/g.

5.4.3 环保生态板材

作者研究团队研究了以竹炭为主要原材料，添加加工辅料、助剂，经过特殊工艺制成炭基复合材料，作为装修和家居的基材，板材具有环保、高强度、强握钉力、防腐防潮、尺寸稳定好等特点，其可替代中密度板、细木工板、强化地板等装修板材，还可进行木质单板、聚氯乙烯（PVC）等贴面，制成炭基—木质装饰材料，可应用于橱柜、家具、地板、踢脚线领域（图 5–7）。

5.4.3 Eco-friendly panel

A man-made panel can be fabricated by extraction of blended bamboo charcoal powder, thermoplastic (like PE, PP, and PVC), and additives, according to study of Author’s group. The bamboo charcoal based panel can be applied as versatile substrate for furniture and decoration, which can be potential replacement to middle density fibreboard (MDF), laminated board, and laminated flooring. The panel features eco-friendliness, high strength and nail-holding power, corrosion-/moisture- resistance, and excellent dimensional stability. It can be PVC- or wood-faced for purpose of decoration or building material applications (Fig 5-7).

图 5–7 竹炭板
Fig 5-7 bamboo charcoal based panel

5.5 其他领域

5.5 Other applications

5.5.1 电容器电极

竹炭可作为双电层电容器的电极材料，目前研究主要集中于竹炭基活性炭的电化学性能，竹炭改性超级电容器电极的制备等方面（图 5–8）。

5.5.1 Capacitor electrode

Bamboo charcoal can be used as electrode material for double layer capacitor, electrochemical behavior of bamboo-based activated carbon is fully explored, and chemically modified bamboo charcoal shows an enormous potential in electrode preparation for super-capacitor (Fig 5-8).

Bamboo-based activated carbon with high specific surface area and specified pore size have been successfully prepared via alkali modification by Author’s group, and studies

张文标、刘洪波、白翔、张勇等研究了竹炭和碱按照不同比例进行混合，制备高比表面积和一定孔径结构的竹质活性炭，主要用作双电层电容器电极，获得较高的比电容。

王力臻等研究了以微波功率640W、辐射时间12min的工艺条件制备竹炭，结果表明该竹炭在100mA/g的电流充放电下，首次放电比电容为242.3F/g，第1000次循环的比电容为229.12F/g，电容保持率为94.56%，并得出竹炭适合大电流充放电的结论。

宁娈等利用竹炭复合二氧化锰（MnO_2），以机械球磨法制得超级电容器，电极比容量可以达到338F/g，100次循环后可以维持在260F/g。

Byoung-Ju Lee 等发现竹质活性炭具有较好的微孔结构，中孔所占比例很大，能够作为离子的快速通道，从而提高电解液的渗透率，比电容要比椰子壳活性炭的比电容高很多，所以可制备出用于超级电容器的活性竹炭。

5.5.2 饲料添加剂

在鸡饲料中添加竹炭可以提高饲料的品质，预防疾病、减少鸡发病率、降低鸡蛋胆固醇、提高产蛋率、改良鸡肉质、减少粪便臭味，改善鸡生长的周围环境；牛饲料中添加竹炭养牛，奶牛的牛奶脂肪增加，味道更加鲜美；用于养猪，能够显著减少养殖周围的气味，改善肉质，强壮内脏功能。

图 5-8 竹炭电容器

Fig 5-8 bamboo charcoal for supercapacitor

demonstrated that an ideal electrochemical performance was obtained in double layer capacitor.

Reported that bamboo charcoal could be a candidate for high current charge-discharge activity. The bamboo charcoal sample prepared by microwave-assistance (640 W, 12 min), showed the specific capacitance reached 243.3 F/g at first charge-discharge cycle, 229.12 F/g at 1000th cycle, with tested current of 100 mA/g.

A study on MnO_2-doped bamboo charcoal prepared by ball milling suggested that its specific capacitance can be as high as 338 F/g, still retaining 260 F/g over 100 cycles in a supercapacitor.

The pore-structured bamboo-based activated carbon had been proved that the permeability of electrolyte enhanced immensely, attributed to mesopores that acting as fast channel for ions, thus specific capacitance of a supercapacitor outweighed that of coconut husk activated carbon.

5.5.2 Feed additive

When bamboo charcoal was added to feed, it has a good chance to prevent poultry

5.5.3 竹炭基肥

魏清泉研究表明，竹炭基肥可部分替代传统化肥，降低化学肥料的施用量，不仅不会影响蔬菜产量，还能提高氮素利用率和改善果实品质；同时还可缓解土壤板结、重金属污染问题，提高肥料吸收利用效率和功能微生物的活性。

5.5.4 复合材料

（1）竹炭纸

竹炭纸是采用特殊工艺将超微竹炭粉与木质纤维结合制成的高分子新型复合材料。产品的可塑性强，能方便地加工成各种保健用品，用后能自然分解或焚烧处理，达到环保要求。具有良好的吸附、除臭、抗菌和屏蔽电磁波等功效，可用于净化空气、高档食品、精密零部件的包装和建筑装潢等领域，还可用于特种行业的特种垃圾处理的包装上，防止垃圾液体的渗出及短时间内变质，吸附垃圾异味。分为单面竹炭纸、夹层竹炭板、双面竹炭纸 3 种。

（2）竹炭皮（革）

竹炭皮（革）又称竹炭复合革（图 5–9），它是利用革基布为底基，涂覆竹炭粉与其他配方原料混合的材料、烘干、压花后所得的产品，是具有皮革特质的环保型纳米竹炭复合材料。“竹炭革”对有害气体（如甲醛、苯）具有良好的吸附性能，可去除环境中的有害气体。

（3）竹炭布

竹炭布有单面、夹层和双面等类型，是以超微竹炭粉为材料，经特殊工艺将竹炭粉涂覆在无纺布或其他基材上而制成的

deceases and morbidity, and provides a less uncomfortable surroundings for animals to accommodate in. Report has that applying bamboo charcoal additive in feeding, egg production, chicken taste and cholesterin level in an egg improved in poultry farming, and meat quality, flavour and fat content of milk also turned out popular in market.

5.5.3 Bamboo charcoal-based fertilizer

According to Wei Qingquan's research, that bamboo charcoal-based fertilizer can partly replace the traditional chemical fertilizer, reduce the amount of chemical fertilizer application, but will not affect the vegetable yield, can also improve the nitrogen use efficiency and improve fruit quality. Also the problems of soil compaction and heavy metal pollution can be alleviated, and the efficiency of fertilizer absorption and utilization and the activity of functional microorganisms can be improved.

5.5.4 Composite

(1) Bamboo charcoal paper

Ultra-fine bamboo charcoal powders was utilized in paper-making, it gave paper an advantage over flexible reprocessing and hazardous disposal aftermath and original properties of bamboo charcoal that continued to hold in the composite. Actually it tends to act as desirable material in decoration, or innovative package material for food or precision instrument and spare part.

(2) Bamboo charcoal leather

Nano-scale bamboo charcoal used as functional additive to coatings that applies to leather substrate, and processed as required, the eco-friendly leather confirmed with a superior adsorption of diminishing toxic and

图 5–9　竹炭皮与竹炭布产品

Fig 5-9　productions of bamboo charcoal leather and fabric

高分子新型复合材料（图 5–9）。其具有良好的吸附分解能力、除臭抗菌能力、远红外线以及防电磁波能力等。可用于制作床垫、鞋垫、口罩、服饰等，起到净化空气、调湿等功效。

harmful gases from air (Fig 5-9).

(3) Bamboo charcoal fabric

A novel polymer composite prepared from ultra-fine bamboo charcoal powders coated non-woven fabric revealed that intrinsic excellence of bamboo charcoal had been consolidated to fabrics, faced fabrics can be applied to manufacture mattresses, masks, cushions, and garments, etc (Fig 5-9).

6 竹提取物

Bamboo Extracts

6.1 定义与分类

6.1.1 竹提取物

（1）定 义

竹提取物指通过物理、化学或生物方法从竹材中提取分离获得的天然产物，其有效成分属于竹子体中的次生代谢产物（糖、蛋白质、脂类和核酸等）。

（2）分 类

竹提取物主要包括竹叶提取物、竹沥、竹醋液、竹笋提取物和竹外皮提取物等。竹叶提取物又包括竹叶黄酮、竹叶多糖、竹叶芳香物、叶绿素、氨基酸、矿物质等。

6.1.2 竹醋液

（1）定 义

竹醋液是竹材或加工剩余物在干馏或炭化的过程中，纤维素、半纤维素等物质受热分解，形成烟雾逸出，经过收集器冷凝收集的具有特殊焦烟味的酸性液体，由竹材所含物质发生化学反应生成，其组分主要是水，还包括酸类、酚类、酮类、醛类、醇类及杂环类等碳氢化合物。竹醋液可分为竹醋液原液（粗竹醋液）、精制竹醋液和蒸馏竹醋液。

（2）分 类

①生产设备：砖土窑、机械窑。

②竹种：散生竹（如毛竹竹醋液）、丛生竹（如麻竹醋液）。

③处理方法：粗竹醋液、精制竹醋液和蒸馏竹醋液。

④用途：农艺、日化、医药用竹醋液（图 6-1）。

6.1 Definition and classification

6.1.1 Bamboo extract

(1) Definition

Bamboo extract is natural product separated from bamboo materials by physical, chemical or biological method, technically it is secondary metabolites (sugars, proteins, lipids and nucleic acids) that makes up the extract.

(2) Classification

The raw materials for the extraction could be bamboo leaves, culms, shoot and processing residues. The bamboo extract can be classified as bamboo sap, bamboo vinegar, bamboo shoot extract, and bamboo leaf extract, and the latter includes bamboo leaf flavonoids, polysaccharides, aromatics, chlorophyll, amino acids and mineral elements.

6.1.2 Bamboo vinegar

(1) Definition

Bamboo vinegar, as known as bamboo pyroligneous liquid, is a liquid product collected from the thermal decomposition of cellulose and hemicellulose during bamboo carbonization or destructive distillation. It consist of water and various organic acids, phenols, ketones, aldehydes, alcohols and heterocyclic hydrocarbons.

(2) Classification

It can be classified on basis of production facilities (kilns or mechanical kilns), raw materials (bamboo species including monopodial and sympodial), refinery processing, and end purposes (gardening, daily chemical, or medicinal).

According to processing, it can be classified as primitive (raw), refined, and

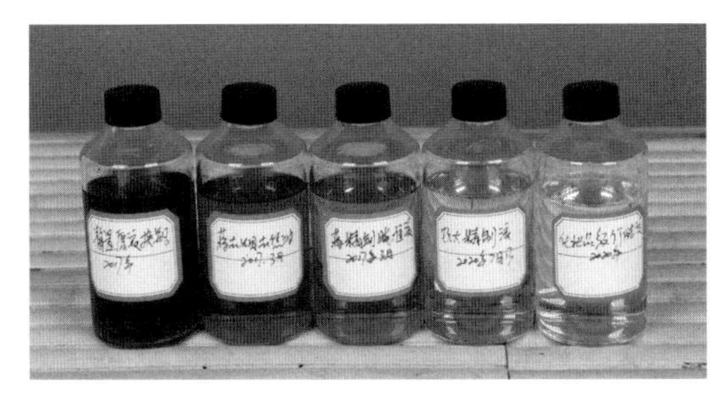

图 6–1 竹醋液

Fig 6-1 bamboo vinegar, refined and raw (L)

a. 竹醋液原液（粗竹醋液）

经过收集器收集的竹醋液，未经任何处理，呈深褐色或红棕色，含有少许炭粒、杂质、悬浮物，粗竹醋液不允许直接使用。

b. 精制竹醋液

精制竹醋液是指竹醋原液通过各种物理和化学方法处理得到的颜色较浅或者无色透明的液体。目前通常有静置法、蒸馏法、过滤法和化学联用等方法。

c. 蒸馏竹醋液

粗竹醋液或精制竹醋液经蒸馏或精馏，除去溶解焦油等高沸点物质后得到的浅黄色或无色透明液体。

6.1.3 竹焦油

竹焦油是竹材热分解的产物之一，呈黑色、黏稠的油状，具有强烈的烟焦气味。

竹焦油的密度略大于水，不（微）溶于水，可溶于汽油等多种有机溶剂，其化学组分复杂，含有大量稠合芳环酚类物质和多种有机物质，如杂酚类、苯系化合物等近百种有机化合物，其中苯并芘等具有致癌性。

竹焦油是有机化工和精细化工原料，是胶粘剂、塑料、染料和药物等合成的重要中间体，还可作为消毒杀菌剂、防腐剂、

distilled bamboo vinegar (Fig 6-1).

a. Primitive (raw) bamboo vinegar

Untreated bamboo vinegar collected from bamboo pyrolysis, is dark brown liquid containing a few bamboo charcoal particles, impurities, and suspended solids, cannot be applied directly.

b. Refined bamboo vinegar

Light colored or transparent clean bamboo vinegar obtained from primitive (raw) bamboo vinegar by physical or chemical treatment. The methods include natural standing, distillation, and filtration-chemical processing.

c. Distilled bamboo vinegar

A yellowish or water clear product purified by removing higher boiling point substances (like soluble tar) by means of distillation of either primitive (raw) bamboo vinegar or refined bamboo vinegar.

6.1.3 Bamboo tar

Bamboo tar is a black, viscous, oil-like substance from bamboo pyrolysis, with a strong burnt smell. It is denser than water, insoluble in water, soluble in solvents. The chemical composition is complicated, containing a large number of polycyclic aromatic phenols, and other organic compounds, of which benzopyrene is carcinogenic to human beings.

Bamboo tar is an important raw material in chemical industry, it can be used as intermediate to synthesis adhesives, plastics, dyes and pharmaceuticals. The application contains germicide, preservative, solvent or fuel, etc.

6.1.4 Fresh bamboo sap

(1) Definition

As a herb of China's traditional medicine, fresh bamboo sap secures an entry in

溶剂和燃料等

6.1.4 鲜竹沥

（1）定　义

鲜竹沥作为传统中药被载入1977年版《中国药典》，为禾本科植物粉绿竹（*Phyllostachys glauca* McClure）、灰竹（*Phyllostachys nuda* McClure）及同属数种植物（淡竹、慈竹、苦竹）的鲜杆经加热后自然沥出的液体，可用于镇咳、祛痰、清肺热等。《本草经集注》记载杨郎生的《中医大辞典》中将竹沥作为中药名。在《全国中草药汇编》《中药大辞典》《中华本草》均有竹沥记载。竹沥别名：竹汁（《本经》），淡竹沥（《别录》），竹油（苏医《中草药手册》）。

（2）指　标

鲜竹沥是一种淡黄色至红棕色的液体，含有如锗、硒、铁、锌等微量元素，具有竹香气、味微甘。

竹沥的pH值：4.0~6.0；

比重：1.008~1.016；

愈创木酚含量：800~1000μg/mL（表6–1）。

Chinese Pharmacopoeia (1977 edition). It is collected from bleeding surfaces of heated freshly-cut culm of common species of gen. *Phyllostachys*, such as *phyllostachys glauca* McClure, *phyllostachys nuda* McClure and *phyllostachys nigra* Munro. It used for therapic purpose of antitussive (prevent or relieve a cough) and expectorant (loosen congestion and help cough up). It is also known as *bamboo juice, bamboo sap* or *bamboo oil*, appeared in multiple classic medicine literature.

(2) Specification

The Specification for fresh bamboo sap can be seen in table 6-1.

(3) Production

① Burning

Burning is the most traditional way to prepare fresh bamboo sap. Newly cut bamboo (culm) with nods removed is heated in the middle by flame, and fresh bamboo sap is collected from the both ends. The production is small and constrained by its low yield and smoke from burning.

② Water extraction

Crush, grind, or slice the bamboo materials, and boil it in hot water or percolate

Table 6-1　Specification for fresh bamboo sap

名称 Item	指标 Description
外观 Appearance	淡黄色至棕色，香气、味甘 Reddish brown to yellowish liquid
气味 Odor and flavor	竹香，甜 Bamboo fragrance, sweetish
pH 值 pH value	4.0~6.0
密度 Specific gravity (density) g/mL	1.008~1.016g/mL
愈创木酚含量 μg/mL Guaiacol (methoxyphenol) content μg/mL	800~1000μg/mL

（3）生产技术

①烧制法

此法为生产鲜竹沥的传统方法。取鲜竹，去两节，中间用火烧之，从竹段两端淌出的液体即为鲜竹沥。但此法收率极低、烟味浓，难以大规模生产。

②水提法

水提法又有煮沸法、渗漉法之分。将鲜竹加工成粉或片，加水煮沸或渗漉提出竹汁。此法收率高于烧制法，适合工业化生产，但因生产中加入了水，稀释了原汁，所以收率还是偏低。

③压榨法

压榨法摈弃了火烤、煮沸、浸渍蒸馏、渗漉等传统方法，透过细胞壁膜成功提取出慈竹最精华的细胞内外汁液——鲜竹汁。工艺路线为：新鲜慈竹切割→清洗→破碎→提取→过滤→脱色→灭菌→罐装。

（4）主要组分

现代科学方法研究发现，鲜竹沥富含多种人体必需氨基酸、黄酮、维生素，以及锗、硒、铁、锌等微量元素，此外还有一些酚类化合物，如愈创木酚。经有机溶剂萃取硅胶、活性炭柱层析等分离得到氨基酸、糖、酚和有机酸。经纸层析和氨基酸自动分析仪检出了丙氨酸、天冬氨酸、谷氨酸、甘氨酸、异亮氨酸、亮氨酸、赖氨酸、甲硫氨酸（蛋氨酸）、脯氨酸、丝氨酸、苏氨酸、酪氨酸、缬氨酸等十三种氨基酸；气相色谱、薄层层析检出有愈创木酚、甲酚、苯酚、甲酸、乙酸等；硅胶柱层分离得三种结晶物，升华纯化后经薄层、红外光谱对照，鉴定结晶为苯甲酸、水杨酸。

to extract bamboo sap, it gives a higher yield than burning method, making a massive production feasible, but the product diluted for water used through the preparation.

③ Cold press

Other than regular *fire and water* methods, cold press have successfully cracked the bamboo cell wall and membrane to collect essence from bamboo. A typical process of primitive bamboo sap production is shown as below:

Bamboo cutting→ Cleaning → Crushing → Extracting → Filtering → De-coloring → Sterilizing → Bottling (Packing).

(4) Constitutes

There are essential amino acids, flavonoids, vitamins, trace elements (Iron, zinc, selenium and germanium), and phenols (guaiacol) were found out in fresh bamboo sap;

Amino acid, organic acid, sugar, and phenols could be separated from bamboo sap by solvent extracted silica gel and activated carbon column chromatography;

There are 13 amino acids, namely, Alanine (Ala), Aspartic acid (Asp), Glutamic acid (Glu), Glycine (Gly), Isoleucine (Ile), Leucine (Leu), Lysine (Lys), Methionine (Met), Proline (Pro), Serine (Ser), Threonine (Thr), Tyrosine (Tyr), and Valine (Val), were detected in bamboo sap by paper chromatography and automatic amino acid analyzer.

Guaiacol (methoxyphenol), cresol, phenol, formic acid, and acetic acid were verified according to gas chromatography (GC) and thin-layer chromatography (TLC) results for bamboo sap analysis.

Purified crystals separated by silica gel column from bamboo sap were confirm to be benzoic acid and salicylic acid (hydroxybenzoic acid) by infrared spectrum investigation.

（5）应用

鲜竹沥是一种药食两用的传统中药，也是目前制作天然绿色功能性饮料的最好原料之一。

①鲜竹汁饮料：以 5 % 鲜竹沥原汁，95 % 饮用水和糖等食用添加剂调配而成，具有解渴、清热解毒、调节肌体等功能。

②药用鲜竹沥：鲜竹沥是传统的中药。

6.1.5 竹叶黄酮

从竹叶中提取出来的具有生理活性的生物黄酮，黄色粉末或晶体，又名竹叶抗氧化剂，英文名 Antioxidant of bamboo leaves，简称 AOB。竹叶黄酮是一种高效多组分协同的生物抗氧化剂，是人体必需的营养素，其成分除了黄酮类化合物以外，还有酚酸、蒽醌、芳香类化合物和锰、锌、锡等微量元素，它们共同构成了竹叶黄酮广泛的生理和药理活性的基础。其中总黄酮糖苷含量≥ 24%，总香豆素类内酯≥ 12%。

6.2 竹醋液生产与加工

6.2.1 生产设备

目前竹醋液是在生产竹炭过程中收集得到的副产品，只需在竹炭生产设备上加装收集装置（图 6–2）。此外，也可采用连续干馏法制备竹醋液。

6.2.2 生产工艺

（1）收集装置

竹醋液收集的装置主要有：冷凝管

(5) Application

Both edible value and medicinal function of bamboo sap was discovered early in China, now it is a novel natural ingredient for functional beverage industrial.

5 % of fresh bamboo sap supplemented water, sugar and food additives can be versatile, nutritious drink in relieving thirst, detoxicating and balancing emotional and physical health, besides, it is also a traditional herbal remedies in traditional Chinese medicine.

6.1.5 Bamboo leaf flavonoid

Bamboo leaf flavonoid, yellow powder or crystalloid, known as antioxidant of bamboo leaves (AOB), is biologically and pharmacologically active flavonoid blend essential for human body. Phenolic acids, anthraquinones, aromatics and some trace elements (manganese, zinc and tin) are also contained in the blend. Total flavonoid glycosides content: ≥ 24 %, total coumarin lactone content: ≥ 12 %.

6.2 Manufacturing and processing of bamboo vinegar

6.2.1 Manufacturing facility

Bamboo vinegar is a acidic byproduct collected during the bamboo carbonization with an identical set of kiln (Fig 6-2).

6.2.2 Production process

(1) Collector

The acid-proof collector is consisted of condensation pipes (bamboo and stainless steel pipe), a condensator or bamboo tar separator (ceramic or stainless steel container), and a receiving container (ceramic, stainless steel, or plastic).

图 6–2 竹醋液收集设备
Fig 6-2 bamboo vinegar collector

（竹片、不锈钢管）、冷凝器或竹焦油分离缸（陶瓷缸、不锈钢）、接收容器（陶瓷缸、不锈钢桶、塑料桶）等。

（2）收集方法

竹醋液的收集温度直接影响竹醋液的质量，不同温度段的竹醋液成分和含量都不相同。砖土窑收集竹醋液主要根据“眼观鼻嗅”方法，即窑温较低时从烟囱中冒出大量水蒸气，不宜收集，水蒸气少了开始收集，当烟囱开始冒青烟时，焦油含量多，结束收集。机械窑生产竹醋液可以根据产品质量要求，选定不同温度段来收集。

（3）保存方法

竹醋液的储存容器需用不锈钢或陶瓷等强耐酸性容器，不能使用铁制容器储存（两者会起反应）。此外，竹醋液应避光保存以免发生光氧化反应，使竹醋液品质改变影响其使用。

6.2.3 产品得率

竹醋液的得率同炭窑类型、炭化速度、收集温度、竹材种类等因素有关。砖土窑

(2) Collection

The quality of bamboo vinegar is mostly affected by its collecting temperature, constitutes and contents varies on temperature intervals. *See and smell* is quite used in a brick kiln case, that is, it is inappropriate to collect when there is a bulk of steam (water vapour) burst out of the chimney, and it is the very right time to start work when the steam drops, again it is not inappropriate to do so when gray smokes pops off, from which the bamboo tar content is climbing. For a mechanic kiln, it is convenient to collect bamboo vinegar in a given temperature range to reach a detailed specification.

(3) Storage

The container for storage should be made of anti-corrosive materials, ironworks are not suitable for storage, otherwise, it will be corroded in that the bamboo vinegar (mainly acetic acid) will react with iron. The products shall be stored with marking or labeling, in a dry, cool places away from sunlight exposure to refrain from photo-oxidation.

6.2.3 Yield

The yield of bamboo vinegar is often affected by kiln style, carbonizing parameters, collecting temperature, and bamboo species. A typical yield for a brick kiln is approx. 8%-10%, and 20%-25% for a mechanic one.

6.2.4 Refining

There are several methods to refine the primitive bamboo vinegar, it shows as follows:

Clarification method:

Simply store bamboo vinegar in a corrosion proof container in the shade, after a long-term standing, upper layer of clarified liquor can be selected as refined bamboo vinegar, and impurity and sediment are

生产时，竹醋液得率一般是竹材 8%~10%，而机械窑一般是竹碎料的 20%~25%。

6.2.4 精制方法

精制竹醋液的方法主要有澄清法、蒸馏法、吸附过滤法、萃取法、连续干馏法等。

澄清法即将竹醋液原液避光保存于耐腐蚀的容器中，经较长时间，杂质、不稳定成分沉淀，上部澄清液即可作为竹醋液使用。

蒸馏法分为常压蒸馏和减压蒸馏，依据各组分沸点不同的特性反复蒸馏，以去除不稳定化合物并溶解焦油，减压蒸馏时，目标组分可在较低温下馏出。

吸附过滤法主要依托活性炭等高吸附能力的吸附剂，除去沉淀物与悬浮物后进行过滤，但存在所需成分被吸附除去的问题。

萃取法可采用不同极性萃取纤维头进行顶空—固相微萃取，该微萃取法属于非溶剂型萃取法，减少样品的分析时间并节省溶剂处理成本，也可以超临界 CO_2 流体为萃取溶剂，以体积浓度为 50%~70% 的乙醇溶液为夹带剂进行萃取，竹醋液精制过程具有快速省时、纯度高、无焦油、无烟熏味等优点。

连续干馏法是在移动床中对竹材干馏进行的平稳炭化，干馏气随时间源源不断地产生并与原料逆向流动（或错流），在加热原料时得到提馏，因而可连续获得初步精制的竹醋液。

discarded from the primitive bamboo vinegar.

Distillation method:

Atmospheric distillation of bamboo vinegar is the process of separating the components from a liquid mixture to diminish unstable compounds and dissolved (soluble) tar by using selective boiling and condensation under atmospheric pressure. A reduced pressure decreases the boiling point of compounds, hence a vacuum distillation method is often conducted to separate target distillates in the primitive bamboo vinegar under reduced pressure and lower temperature.

Adsorption-filtration method:

With removal of sediment and suspended substance in the primitive bamboo vinegar by absorption of activated carbon, the bamboo vinegar is treated with filtering to obtain refined product. Note that necessary ingredients may be disposed of.

Continuous distillation:

Gentle carbonizing of bamboo materials is occurred in destructive distillation in a moving bed, as carbonization goes, distillate gases released in a counter-flow of raw materials are spontaneously heated to separated, ultimately initially purified bamboo vinegar is attained from continuous distillation.

Extraction method:

Headspace solid phase microextraction (SPME) method is a novel non-solvent extraction using SPME fibers with diverse polarity to refine bamboo vinegar, it can reduce both analysis time span and chemical cost.

In addition, using super-critical fluid (CO_2) as extractive solvent, ethanol (50%-70% by vol.) as entrainer (cosolvent), refined bamboo vinegar can be quickly obtained featuring high concentration, and tar-/smoke smell-free.

6.3 竹醋液成分与指标

6.3.1 成 分

竹醋液组分采用气相色谱—质谱联用仪（GC-MS）测试。它是一种由水（80 %左右）、有机酸、醛类、氨基酸类、酚类、酮类、醇类及酯类等200多种组分组成的混合物。

竹醋液组分含量的形成与竹材的化学组成密切相关，其组分随竹材的种类、质量、竹材的炭化设备、炭化工艺条件、收集方法、处理工艺、贮存等不同而有所差异。

不同竹种制得的竹醋液组分含量、有机物种类各不相同，主要成分是以乙酸为主的有机酸。

生产设备对竹醋液的品质至关重要，采用机械窑生产时的竹醋液品质比较稳定；采用砖土窑生产的竹醋液品质比较难控制，但有机酸含量高，其中醋酸可达7.0%。

不同温度收集的竹醋液化学组分中，各类物质和含量各不相同，有机酸类物质的含量差别最明显。

精制竹醋液的有机酸含量比竹醋液原液高且精制竹醋液的有机酸种类也比竹醋液原液多，而竹醋液原液的酚类、醛类、酮类、酯类分别比精制竹醋液高且醇类和其他有机物含量远高于精制竹醋液；蒸馏的竹醋液的有机酸类物质减少，无酚类物质，但蒸馏竹醋液放置4个月后有机酸物质，尤其醛类物质又出现，醇类物质也比原液少。活性炭过滤的竹醋液，酚类等其他大分子物质较少，表明活性炭具有强的

6.3 Constituents and properties of bamboo vinegar

6.3.1 Constituents

Bamboo vinegar is a complicated blend consisted of hundreds of chemicals, including water, organic acids, aldehydes, amino acids, phenols, ketones, alcohols and esters according to a GC-MS analysis. The content depends on chemical components of bamboo materials. Constituents of bamboo vinegar change with bamboo species and quality, carbonization equipment and parameters, collecting and processing method, and storage conditions, respectively, nevertheless, acetic acid is the leading chemical in the bamboo vinegar.

Bamboo vinegar quality is mostly affected by carbonization equipment. It produces consistent quality product by a mechanical kiln; while produced by brick kiln, its quality is not as stable as that of its counterpart, but with a higher concentration of organic acid, for instance, the acetic acid content could be up to 7.0 %. Constituents and contents of bamboo vinegar vary on carbonization temperature, amongst which the organic acid content differs remarkably.

Refined bamboo vinegar show a higher content of organic acid and complex compositions than that of primitive (raw) bamboo vinegar, and the rest organic compounds, alcohols in particular, content is much lower.

Distilled bamboo vinegar confirmed with less organic acids, and little phenols, but aldehydes may be found after a 4-month storage. Activated carbon-filtered bamboo vinegar appears light colored and low level of larger molecules, like aldehydes. The quality is generally influenced by package, storage place and time span, sometimes even samples taken

吸附性能，吸附竹醋液中颜色深的大分子物质，颜色变浅。

竹醋液储存时，容器材料、存放时间、存放地点、取液部位等也将直接影响他的品质，极易受光照、氧化等因素影响造成内部组分变化。

6.3.2 基本理化指标

竹醋液的基本性质包括外观、气味、密度、pH 值、有机酸含量、可溶解焦油含量和折光率等。竹醋液的基本指标见表 6–2。

6.3.3 指标分析方法

（1）竹醋液的有机酸含量（以醋酸计）

竹醋液中的有机酸以醋酸为主，通常以醋酸含量来表示有机酸含量，一般要求 ≥ 3.0%。

from different spot in an identical container are founded with content variation.

6.3.2 General properties

Basic properties, including appearance, odor, transparency, density, acidity, water, organic acids, and soluble tar content, are summed up in table 6-2:

6.3.3 Analysis method

(1) Determination of organic acids content (as acetic acid)

Transfer 2 mL of sample to a weighed conical flask, and precisely weigh it, add 60 mL newly-boiled deionized water and droplets of phenolphthalein indicator, titrate the solution with 1N sodium hydroxide (NaOH) standard solution, a slight persisting pink color signals the endpoint of the titration, and record the volume of standard solution consumed,

表 6–2 竹醋液的理化指标

Table 6-2 Typical properties of bamboo vinegar

项目 Item	指标 Typical description
外观 Appearance	深红褐色透明液体 Dark brown clean liquid
气味 Odor	特殊烟焦味 Special smoke smell
密度 Density	1.010~1.050 g/cm^3
pH 值 pH value	2.4~3.2
有机酸（以醋酸计） Organic acids content（as acetic acid）	≥ 4.0 %
可溶解焦油 soluble tar content	≤ 1.0 %
折光率 Brix value	≥ 4.0

测定方法：取试样约 2mL，放入已知重量的锥形瓶中，精确称量，加新煮沸过的冷蒸馏水 60mL，加酚酞指示剂 3 滴，用 1N 的氢氧化钠（NaOH）滴定至微红色。醋酸含量计算公式如下：

醋酸含量 $=V\times C\times 0.06005/m\ \times 100$（%）

（式 6-1）

式中：V——耗用氢氧化钠（NaOH）标准液毫升数，mL；

C——氢氧化钠（NaOH）标准液的当量浓度，mol/L；

m——试样质量，g。

（2）pH 值

竹醋液显示强酸性。常用方法是使用 pH 试纸、pH 计或 pH 测定仪。生产上常用 pH 计。

（3）密　度

测定时预先将试样温度调至 20℃ ± 1℃后（此温度保持到测定结束），将试样沿玻璃棒慢慢地注入清洁干燥的量筒中，不得使试样产生气泡和泡沫，拿住波美计（量程：1.000~1.100g/mL）上端，将其慢慢地放入试样中，注意不要接触筒壁。当波美计在试样中停止摆动后，即记下液面与刻度线交接处的数据精确到 0.001。平行测定三次，测定结果之差不超过 0.002，计算平均值，即为试样比重，单位为 g/cm^3。

（4）竹醋液的溶解焦油含量

将竹醋液放入蒸发皿，置于 125 ± 5℃烘箱内，烘 8h，其重量与原竹醋液重量的比值即为竹醋液溶解焦油含量。这个数值越小说明竹醋液纯度越高、品质也越好。

（5）竹醋液的折光率

折光式糖度计先以纯水校正零点，然

then calcalate the organic acids content by formula (6-1)

Organic acids content (as acetic acid)
$=V\times C\times M/\mathrm{m}\times 100\%$ (6-1)

Where,

V refers to the volume of NaOH standard solution consumed expressed in milliliters;

C refers to the concentration of the NaOH standard solution expressed in mole/L;

M refers to the molecular weight of acetic acid, which is 60.05;

m refer to the weight of titrated bamboo vinegar expressed in grams.

(2) Determination of acidity (pH value)

Acidity of bamboo vinegar can be expressed by pH value, which is technically measured by indicator paper or meter.

(3) Determination of density

A Baumé hydrometer is a tool used to find a solution’s specific gravity, as well as to estimate its concentration. It looks like a thermometer with a bulb on one end and degrees marked along its length. After placing the hydrometer (range: 1.000-1.100 g/mL) in the bamboo vinegar (which can be loaded into a measuring cylinder), the degrees’ reading corresponds to one of two different scales for sample with higher densities than water (bamboo vinegar in this case). Most modern Baumé hydrometers are calibrated to provide accurate readings at 20℃. The best practice for using the hydrometer and preserving its integrity is to heat or cool sample liquid to the correct temperature before measuring it. If that’s not possible, however, reading a chart or use an online calculator to find the Baumé temperature correction according to the conditions when recording data.

(4) Determination of soluble tar content

后用擦镜纸将折光棱镜镜面拭净，滴上竹醋液试样数滴，合上盖板，使溶液均布于镜面，将仪器进光窗对准光源或明亮处，调节目镜的视度圈，使视场内明暗分界线清晰可见，读取分界线所对准的刻度值。要求平行测定 3 次，相对标准偏差小于 3%。

6.4 竹醋液应用

6.4.1 功 效

竹醋液是多成分组成的混合物，主要由有机酸、酚类、醇类、酯类、酮类、杂环类等物质组成，具有杀菌、抑菌、消毒、除臭、抗氧化、促进植物生长、防腐、驱虫等功效。

6.4.2 产 品

竹醋液消毒、除臭、抑菌系列（如工业用除臭剂、日用除臭剂、土壤消毒剂、医用杀毒剂、种子消毒剂、环境除臭剂）。

竹醋洗涤洁肤系列（如竹醋液洗发液、竹醋液沐浴露、竹醋液洗面奶、竹醋液牙膏）（图 6–3）。

竹醋液添加剂系列（如叶面肥、饲料添加剂、防腐剂）。

竹醋粉系列（如足贴）。

6.5 其 他

6.5.1 贮 存

竹醋液是一种强酸性物质，不稳定，

Transfer 5-20 mL of sample to a weighed pre-cleaned evaporation dish, and precisely weigh it, oven heat it at 125℃ for 8 h. The difference in weight before and after the heating is the percentage of the soluble tar content.

(5) Determination of brix value

A refractometer(saccharimeter, brixmeter) is often used to determine the brix value of bamboo vinegar. Before determinaton, zeroing an optical refractometer with the knob or screw. Note that pure water reads at zero, adjust it as necessary. Open the daylight plate and wipe the prism of a refractometer, add a few drops of bamboo vinegar on the prism using a pipette, close the daylight plate so that the liquid spreads evenly across the surface of the prism. Hold the refractometer horizontally in a bright light source, when first look into it, the numbers may be blurry, turn the eyepiece until the numbers come into focus, record the number which is the brix value of the tested bamboo vinegar.

6.4 Applications of bamboo vinegar

6.4.1 Function

Bamboo vinegar is a multicomponent product collected from bamboo carbonizing, which consists of organic acids, phenols, alcohols, esters, ketones and heterocyclic compounds, featuring sterilization, bacteria resistance, deodorization, anti-oxidation, plant growth-promoting, and insect repelling.

6.4.2 Product

Bamboo vinegar can be applied as sterilizer, deodorizer, and bactericide both in industrial and at home. It can be a special ingredient added to natural fertilizer, feed,

图 6–3 竹醋液洗护产品

Fig 6-3 bamboo vinegar additive in personal care products

长时间放置会出现沉淀，见光易发生光化学反应，组成成分发生改变，颜色也会变深。因此要注意存放容器、时间和地点；保存竹醋液应采用不透明的、耐酸容器，避免强光照射、雨淋。

6.5.2 食 用

虽然竹醋液是纯天然的产品，但也含有少量对人体有危害物质，如酚类物质、焦油及重金属等，不可随意饮用。

目前日本已有可直接饮用的竹醋液，是经过二次以上的蒸馏法再精制成的产品，作为可食用竹醋液产品要符合可食用安全的相关标准。

6.5.3 外 敷

一般而言，竹醋液可直接接触皮肤，但因个体体质、皮肤等特殊性，建议使用前先试滴在手背上，若感觉不适或有过敏现象则停止使用。作为外敷的竹醋液需要精制处理并且产品指标达到有关标准要求。另外，使用时请避开伤口部位。

and personal care products, such as shampoo, toothpaste and cleanser, etc (Fig 6-3).

Bamboo vinegar is particularly effective for eliminating mould and insect infestations in storage areas, textiles or mats. It also used as a fertiliser for treating plants with infested roots or leaves by simply adding some when watering. Just like bamboo charcoal, it can help extinguish unpleasant odours at home and in specific circumstance.

6.5 Other information

6.5.1 Storage

Bamboo vinegar is of strong acidity, and easy to photo-degrade, the bamboo vinegar product shall be packed in acid-proof, non-transparent container, stored with marking or labeling in a dry, cool place to avoid moisture and light exposure, sediment will occur after a long term storage and the color will be darker caused by the photo-oxidation.

6.5.2 Edible

Being a natural product, bamboo vinegar still contains harmful chemicals, such as phenols, tar and heavy metals. Refined bamboo vinegar prepared by re-distillation could be consumable, nevertheless, it should reach specific safety requirements or mandatory standards.

6.5.3 External application

Generally it does not matter when bamboo vinegar directly used in skin contact. Only after a skin test did one can use it for purpose of external application, and wounds do not apply. Bamboo vinegar for skin care should meet specific medical and safety requirements.

6.5.4 标 准

中国：GB/T 31734—2015《竹醋液》；

2005 年日本竹炭、竹醋液生产者协会发布的《木醋液 · 竹醋液的规格》《竹醋液品质》标准。

韩国食品药品安全厅发布《竹（木）醋液天然添加物许可》标准。

6.5.4 Standard

China: GB/T 31734—2015 Bamboo pyroligneous liquid;

Japan: Specifications of Wood Vinegar & Bamboo Vinegar and quality of bamboo vinegar. (Issued by Japan Banboo charcoat and Bamboo Vinegar Associtiion).

South Korea: Bamboo Vinegar and wood Vinegar Liquid Natural Additives Permitted Baseline (Issued by the South Korea Food and Dry Safety Agency).

参考文献
Reference

[1] 姜在允 . 木炭拯救性命：徐徐揭开的秘密 [M]. 金莲兰，译 . 武汉：中国地质大学出版社，2004，1–254.

JIANG Z Y,JIN L L.Charcoal Saves Lives : Secrets Slowly Revealed[M].Wuhan :China University of Geosciences Press,2004，1–254.

[2] 张文标，李文珠，张宏 . 竹炭 . 竹醋液的生产与应用 [M]. 北京 : 中国林业出版社 .2006，1–168.

ZHANG W B,LI W Z,ZHANG H.Production and Application of Bamboo Vinegar[M].Beijing: China Forestry Publishing House.2006,1–168.

[3] 王静波，陈文照，张文标，等 . 樱花炭语 [M]. 杭州：浙江文艺出版社，2006,1–183.

WANG J B,CHEN W Z,ZHANG W B,et al.The Cherry Blossom Charcoal Language[M]. Hangzhou :Zhe Jiang LIterature and Art Publishing House,2006,1–183

[4] 江泽慧 , 张东升 , 费本华 , 等 . 炭化温度对竹炭微观结构及电性能的影响 [J]. 新型炭材料 ,2004(4):249–253.

JIANG Z H,ZHANG D S,FEI B H, et al.Effect of Carbonization Temperature on the Mmicrostructure and Conductivity of Bamboo Charcoal[J].New Carbon Material,2004(4):249–253.

[5] 杨丽 , 刘洪波 , 张东升 , 等 . 竹炭微观结构的电子显微学研究 [J]. 电子显微学报 , 2011, 30(2): 137–142.

YANG L,LIU H,ZHANG D,et al.Electron microscopy Study on Microstructure of Bamboo Charcoal [J].Journal of Chinese Electron Microscopy Society,2011,30(2):137–142.

[6] 卢克阳 , 张齐生 , 蒋身学 . 竹炭吸湿性能的初步研究 [J]. 木材工业 ,2006(3):20–22.

LU K Y,ZHANG Q S,JIANG S X.A Primary Study of Bamboo Charcoal Moisture Absorption[J]. China Wood Industry,2006(3):20–22.

[7] Liao P, Ismael Z M, Zhang W, et al.Adsorption of Dyes from Aqueous Solutions by Microwave Modified Bamboo charcoal[J]. Chemical Engineering Journal,2012, 195: 339–346.

[8] 周建斌 , 邓丛静 , 傅金和 , 等 . 竹炭负载纳米 TiO2 吸附与降解甲苯的研究 [J]. 新型炭材料 , 2009, 24(2): 131–135.

ZHOU J B,DENG C J,FU J H,et al.Adsorption and Degradation of Toluene by nano–TiO2 Supported on Bamboo Charcoal[J].New Carbon Material,2009,24(2): 131–135.

[9] Saito Y , Mori M , Shida S , et al.Formaldehyde Adsorption and Desorption Properties of Wood–

Based Desorption Properties of Charcoal[J]. journal of the Japan Wood Researh Society,2000, 46 (6): 59–61.

[10] Asada T , Ishihara S , Yamane T , et al.Science of Bamboo Charcoal: Study on Carbonizing Temperature of
Bamboo Charcoal and Removal Capability of Harmful Gases[J].Journal of Health Science,2002,48(6):473–479.

[11] 徐亦钢 , 石利利 . 竹炭对 2,4- 二氯苯酚的吸附特性及影响因素研究 [J]. 生态与农村环境学报 ,2002,018(001):35–37.
XU Y G,SHI L L.Adsorption Characteristics and Influencing Factors of 2,4–Dichlorophenol on Bamboo Charcoal[J].Journal of Ecology and Rural Environment,2002,018(001):35–37.

[12] 肖继波 , 陈斌 , 曹玉成 . 竹炭对染料的吸附性能研究 [J]. 福建林业科技 ,2006(04):117–120+127.
XIAO J B,CHEN B,CAO Y C.Study on the Adsorption Properties of Bamboo Charcoal for Dyes[J].Fujian Forestry Science and Technology,2006(04):117–120+127.

[13] 杨磊 , 陈清松 , 赖寿莲 , 等 . 竹炭对甲醛的吸附性能研究 [J]. 林产化学与工业 , 2005, 25(001):77–80.
YANG L,CHEN Q S,LAI S L,et al.Study on Formaldehyde Adsorption Performance of Bamboo Charcoal [J].Forest products Chemistry and Industry,2005, 25(001):77–80.

[14] 钟雨婷 , 贾贞超 , 闫九明 , 等 . 食用竹炭粉的急性毒性与致突变性研究 [J]. 检测研究 ,2015, 27(2): 142–145.
ZHONG Y T,JIA Z C,YAN J M,et al.Study of Acute Toxicity and Mutagenicity of Bamboo Charcoal Powder.[J].Test Research,2015, 27(2): 142–145.

[15] 冯初国，陈顺伟，徐叨兴. 竹炭对茶叶贮藏品质的动态分析初报 [J]. 浙江林业科技，2005,25(3): 15–17.
FENG C G,CHEN S W,XU D X.Preliminary Report on Dynamic Analysis of Bamboo Charcoal on Tea Storage Quality[J].Zhejiang Forestry Science and Technology,2005,25(3)：15–17.

[16] 闫鸿敏 , 王朝生 , 邹瑜 , 等 . 竹炭纤维的开发与应用 [J]. 针织工业 , 2007 (3): 13–16.
YAN H M,WANG C S,ZOU Y,et al.Development and Application of Bamboo Charcoal Fiber[J].Knitting Industry,2007 (3): 13–16.

[17] Wu K H, Ting T H, Wang G P, et al.Synthesis and Microwave Electromagnetic Characteristics of Bamboo
Charcoal/Polyaniline Composites in 2 – 40 GHz[J]. Synthetic Metals,2008, 158(17–18): 688–694.

[18] Lin J H, Chen A P, Lin C M, et al.Manufacture Technique and Electrical Properties Evaluation

of Bamboo Charcoal Polyester/Sainless Steel Complex yarn and Knitted Fabrics[J].Fibers and Polymers,2010, 11(6): 856–860.

[19] 王祝来，陈玉峰，周琼，等．竹炭对水溶液中氨氮的吸附特性研究 [J]. 工业用水与废水，2009(4):70–73.

WANG Z L,CHEN Y F,ZHOU Q. et al.Adsorption Characteristics of Bamboo Charcoal on Ammonia Nitrogen in Aqueous solution[J]. Industrial Water and Waste Water,2009(4):70–73.

[20] 张启伟，王桂仙．竹炭对溶液中铅（Ⅱ）离子的吸附行为研究 [J]. 丽水学院学报，2005, 27(5): 60–63.

ZHANG Q W,WANG G X.Adsorption of pb（ Ⅱ ）ions in Solution by Bamboo Charcoal[J]. Journal of Lishui University,2005, 27(5): 60–63.

[21] 刘创，赵松林，许坚．竹炭对水溶液中 Cd(Ⅱ) 的吸附研究 [J]. 科学技术与工程，2009, 9(011):3009–3012.

LIU C,ZHAO S L,XU J.Adsorption of Cd(Ⅱ) from Aqueous Solution by Bamboo Charcoal[J]. Science, Technology and Engineering,2009, 9(011):3009–3012.

[22] 韦冬芳，韦仲华，金城风鹤，等．竹炭陶的制备及其气体吸附和调湿性能 [J]. 林业工程学报，2020, 005(001):P.109–113.

WEI D F,WEI Z H,JIN C F H,et al.Preparation of Bamboo Charcoal Pottery and Its Gas Adsorption and Hygroscopic Properties[J].Journal of For–Etry Engineering, 2020, 005(001):P.109–113.

[23] 彭虎，王平，匡猛，等．竹炭笼芯紫砂毫 / 微球空气净化器制备与性能研究 [J]. 中国陶瓷，2016(5):64–69.

PENG H,WANG P,KUANG M,et al.Preparation and Performance of Bamboo Charcoal Cage Core purple sand Millibead/Microsphere Air Purifier[J].Chinese Ceramic, 2016(5):64–69.

[24] 王喜华，陈港．纳米 TiO_2/ 竹炭涂布纸对甲醛吸附降解性能的研究 [J]. 中国造纸，2010(10):11–15.

WANG X H,CHEN G.Study on Formaldehyde Adsorption and Degradation Properties of Nano–TiO_2/Bamboo Charcoal Coated Paper[J].China's Paper Industry,2010(10):11–15.

[25] 张文标，林启晨，徐冲霄，等．我国竹炭研究现状和展望 [J]. 竹子研究汇刊 ,2014,33(1):1–6.

ZHANG W B,LIN Q C,XU C X, et al.Present Situation and Prospect of Bamboo Charcoal Study in China[J].Journal of Bamboo Research,2014,33(1):1–6.

[26] 沈跃华，吴来明，解玉林，等．纤维调湿板，CN101343850[P].

SHEN Y H,WU L M,JIE Y L,et al.Fiber Hygroscopic Board, CN101343850[P].

[27] 李文彦，杨辉，郭兴忠，等．用竹炭和硅藻土为原料制备含炭建筑材料 [J]. 材料科学与工程学报，2011, 29(1): 7–11.

LI W Y,YANG H,GUO X Z,et al.Bamboo Charcoal and Diatomite Were Used as Raw Materials to Prepare Carbon – Containing Building Materials[J].Journal of Materials Science and Engineering,2011,29(1): 7–11.

[28] 吉行那惠子，原田志津男，國府俊則，奥野守人，黒岩朱美．廃ガテスを用いた竹炭複合せテミツケスの開発 [C]. 日本建筑仕上学会，2006，207–210.

[29] Gabhi R S, Kirk D W, Jia C Q.Preliminary Investigation of Electrical Conductivity of Monolithic Biochar[J]. Carbon, 2017, 116: 435–442.

[30] 马建伟 , 王慧 , 罗启仕 . 电动力学 – 新型竹炭联合作用下土壤镉的迁移吸附及其机理 [J]. 环境科学 , 2007, 28(8): 1829–1834.

MA J W,WANG H, LUO Q S.Migration and Adsorption Mechanism of Cadmium in Soil Under the Combined Action of Electrodynamics and New Bamboo Charcoal[J].Environmental Sciences,2007,28(8): 1829–1834.

[31] 杨磊 . 竹炭基功能型复合板材的研制 [J]. 莆田学院学报 , 2006, 13(2): 77–80.

YANG L.Development of Bamboo Charcoal Based Functional Composite Plate[J].Journal of Putian University,2006, 13(2): 77–80.

[32] 刘洪波 , 常俊玲 , 张红波 , 等 . 竹炭基高比表面积活性炭电极材料的研究 [J]. 炭素技术 , 2003 (5): 1–7.

LIU H B,CHANG J L,ZHANG H B,et al.Study on Bamboo Charcoal Based Activated Carbon Electrode Material with High Specific Surface Area[J].Carbon Technology,2003 (5): 1–7.

[33] 王力臻 , 方华 , 张爱勤 , 等 . 微波辐射法制备竹炭电极材料 [J]. 电池 , 2010, 40(2): 77–79.

WANG L Z,FANG H, ZHANG A Q,et al.Bamboo Carbon Electrode Material was Prepared by Microwave Radiation Method[J].Battery,2010, 40(2): 77–79.

[34] 宁娈 , 陈晓红 , 宋怀河 , 等 . MnO_2/ 竹炭超级电容器电极材料的性能 [J]. 北京化工大学学报 (自然科学版), 2010, 37(2): 65.

NING L,CHEN X H,SONG H H,et al.Performance of MnO_2/ Bamboo Charcoal Supercapacitor Electrode Material[J].Journal of Beijing University of Chemical Technology (Natural Science Edition),2010, 37(2): 65.

[35] Lee B J, Kim Y J, Horie Y, et al. Bamboo–Based Activated Carbons as an Electrode Material for Electric Double layer Capacitors (EDLCs)[J].Carbon, 2004, 4: 1–4.

[36] 魏清泉 , 史元元 , 董民 , 等 . 竹炭基肥与猪厩肥混施对生菜产量与品质的影响 [J]. 蔬菜 , 2016 (1): 4–6.

WEI Q Q,SHI Y Y,DONG M,et al.Effects of Mixed Application of Bamboo Charcoal Fertilizer and Pig Manureon Yield and Quality of Lettuce[J].Vegetable,2016 (1): 4–6.